The International Library of Bioethics

Volume 83

The *International Library of Bioethics* – formerly known as the International Library of Ethics, Law and the New Medicine comprises volumes with an international and interdisciplinary focus on foundational and applied issues in bioethics. With this renewal of a successful series we aim to meet the challenge of our time: how to direct biotechnology to human and other living things' ends, how to deal with changed values in the areas of religion, society, and culture, and how to formulate a new way of thinking, a new bioethics.

The *International Library of Bioethics* focuses on the role of bioethics against the background of increasing globalization and interdependency of the world's cultures and governments, with mutual influencing occurring throughout the world in all fields. The series will continue to focus on perennial issues of aging, mental health, preventive medicine, medical research issues, end of life, biolaw, and other areas of bioethics, whilst expanding into other current and future topics.

We welcome book proposals representing the broad interest of this series' interdisciplinary and international focus. We especially encourage proposals addressing aspects of changes in biological and medical research and clinical health care, health policy, medical and biotechnology, and other applied ethical areas involving living things, with an emphasis on those interventions and alterations that force us to re-examine foundational issues.

More information about this series at http://www.springer.com/series/16538

Michael Kühler · Veselin L. Mitrović
Editors

Theories of the Self and Autonomy in Medical Ethics

Springer

Editors
Michael Kühler
Department of Philosophy
University of Twente
Enschede, Overijssel, The Netherlands

Veselin L. Mitrović
Institute of Social Sciences
University of Belgrade
Belgrade, Serbia

ISSN 2662-9186 ISSN 2662-9194 (electronic)
The International Library of Bioethics
ISBN 978-3-030-56702-6 ISBN 978-3-030-56703-3 (eBook)
https://doi.org/10.1007/978-3-030-56703-3

This Springer imprint is published by the registered company Springer Nature Switzerland AG
The registered company address is: Gewerbestrasse 11, 6330 Cham, Switzerland

Preface

This volume is the result of a longstanding cooperation between the editors and some of the contributors. Starting with a workshop on "Issues in Theoretical and Applied Ethics," organized by Gunnar Scott Reinbacher and Jörg Zeller in 2011 in Klitgaarden, Denmark, the central topic of our shared discussion since then had quickly been found, namely, the controversial relation between debates and arguments in theoretical ethics and metaethics, on the one hand, and applied ethics, esp. medical ethics, on the other hand. Since then an edited volume based on this first workshop followed (Zeller, Jörg/Riis, Ole Preben/Nykänen, Hannes (eds.): *Issues in Theoretical and Applied Ethics*, Aalborg: Aalborg University Press) as well as two further workshops: "Applied Ethics and Applying Ethics," organized by Michael Kühler and Jörg Zeller in 2013 in Münster, Germany, and "Theories of the Self and Respect for Autonomy in Palliative Care and End-of-Life Decisions," organized by Veselin Mitrović and Michael Kühler in 2016 in Belgrade, Serbia. The latter workshop has been the starting point for the current volume, which contains a number of revised contributions to this workshop but also a number of additional contributions by other colleagues, thus joining our ongoing discussion.

As with all such volumes, they are the result of many people's contributions and help. First of all, we would like to express our utmost gratitude to all contributors for putting in so much hard work to provide this volume with so many excellent and thought-provoking chapters. Furthermore, we would like to say a special "thank you" to Cecil Joselin Simon, Christopher Wilby, Floor Oosting, and Cynthia Kroonen at Springer for their tremendous support and truly admirable patience when it came to putting this volume together. Last but certainly not least, we are very grateful to Lucie White and Rachel Fedock for their invaluable help in proofreading and language editing. We cannot stress enough how much we appreciate their support.

Belgrade, Serbia
Enschede, The Netherlands
June 2020

Veselin L. Mitrović
Michael Kühler

Contents

About the Editors

Michael Kühler is Assistant Professor at the University of Twente, Netherlands and "Privatdozent" (roughly equaling Associate Professor) at Münster University, Germany. His research interests include ethics, medical ethics, metaethics, and political philosophy. His recent publications include "Technological Moral Luck," in: Beck, Birgit/Kühler, Michael (eds.): *Technology, Anthropology, and Dimensions of Responsibility*, Techno:Phil 1, Stuttgart: Metzler, 2020, 115–132, https://doi.org/10.1007/978-3-476-04896-7_9, "Romantische Liebe und die Freiheit zu gehen," in: Dietz, Simone/Foth, Hannes/Wiertz, Svenja (eds.): *Die Freiheit zu gehen. Ausstiegsoptionen in politischen, sozialen und existenziellen Kontexten*, Wiesbaden: Springer, 2019, 195–222, "*Modus Vivendi* and Toleration," in: Horton, John/Westphal, Manon/Willems, Ulrich (eds.): *The Political Theory of Modus Vivendi*, Dordrecht: Springer, 2019, 235–253, "Toleranz und/oder Paternalismus im engeren sozialen Nahbereich?," in: *Zeitschrift für Praktische Philosophie* 4, 2017, 63–86; and "What if I Cannot Do What I Have to Do? Notions of Personal Practical Necessity and the Principle 'Ought Implies Can'," in: Bauer, Katharina/Mieth, Corinna/Varga, Somogy (eds.): *Dimensions of Practical Necessity. "Here I Stand. I Can Do No Other,"* Houndmills: Palgrave Macmillan, 2017, 87–107

Veselin L. Mitrović is a senior research associate at the Institute of Social Sciences at the University of Belgrade. His research activities feature topics in Bioethics, Disaster Study, Sociology of Culture and Social Action in different strata (from marginal to elite). His recent publications include "Ethics and Floods: A Systematic Review." *Disaster Medicine and Public Health Preparedness.* 13/4, 2019, 817–828. "Parent's Religious and Secular Perspectives on IVF Planning in Serbia," *Journal for the Study of Religions and Ideologies* 15/43, Spring 2016, 48–81; "From Global Bioethics to Neuroethics," *TEME* XL/4, 2016, 1457–1475 (Serbian); "Resilience: detecting vulnerability in marginal groups," *Disaster Prevention and Management* 24/2, 2015, 185–200; "The Contingency of the 'Enhancement' Arguments: The Possible Transition From Ethical Debate to Social and Political Programs," *Journal for the Study of Religions and Ideologies* 13/37, Spring 2014, 93–124; "Ethos of the Elite and Future of Serbia," *Balkanske sinteze* 1/1, 2014, 167–189 (Serbian); "Nanotechnologies and Global Survival," *Sociologija* LIV/1, 2012, 37–54 (Serbian);

"The Myth of Moral Enhancement: Back to the Future?," *Filozofija i društvo* XXIII/2, 2012, 111–123 (Serbian); "Human Enhancement: Toward the Creation of Patterns of Injustice?" Zeller, Jörg/Riis, Ole Preben/Nykänen, Hannes (eds.): *Issues in Theoretical and Applied Ethics*, Aalborg: Aalborg University Press, 2013; "Study," *Sociology in Serbia 1959–2009, Institutional Development*, 59–107, (ed.) Marija Bogdanović, (Beograd: Službeni glasnik i Filozofski fakultet Univerzitet u Beogradu, 2009) (Serbian). He is also author of the following books written in Serbian: *Apathetic Society* (2015); *The Stride of Bioethics, New Bio-Technologies and Social Aspects of the "Enhancement" of the Healthy* (2012) and *Jazz as Socio-Cultural Improvisation – A Qualitative Research of Social Mobility* (2012).

Chapter 1
Introduction

Michael Kühler and Veselin Mitrović

1.1 Autonomy and the Self in Recent Medical Ethics

For the last few decades, the principle of respecting patients' autonomy has been of major importance in medical ethics, notably in terms of the requirement of gaining patients' informed consent for any treatment to be considered legitimate.[1] However, there is still an ongoing debate about its exact understanding and implications. It is not only a matter of dispute how to analyze the concept of personal autonomy in general,[2] but also whether and to what degree resulting specific conceptions of autonomy are suitable for being used in medical ethics in particular.[3]

---

[1]Cf., for example, Eyal (2012) and Article 5, "Autonomy and individual responsibility," of the *Universal Declaration on Bioethics and Human Rights*: "The autonomy of persons to make decisions, while taking responsibility for those decisions and respecting the autonomy of others, is to be respected. For persons who are not capable of exercising autonomy, special measures are to be taken to protect their rights and interests" (UNESCO 2005, 77).

[2]See, for example the contributions in Mackenzie and Stoljar (2000), Christman and Anderson (2005), Taylor (2005); and Kühler and Jelinek (2013a). For current overviews of the debate, see Buss (2013), Christman (2015), and Stoljar (2015).

[3]See, for example Maclean (2009).

M. Kühler (✉)
University of Münster, Münster, Germany
e-mail: michael.kuehler@uni-muenster.de

University of Twente, Enschede, The Netherlands

V. Mitrović
University of Belgrade, Belgrade, Serbia
e-mail: mitrove@gmail.com

M. Kühler and V. L. Mitrović (eds.), *Theories of the Self and Autonomy in Medical Ethics*, The International Library of Bioethics 83,
https://doi.org/10.1007/978-3-030-56703-3_1

In this regard, Tom Beauchamp and James Childress, in their highly influential *Principles of Biomedical Ethics*,[4] start their analysis of autonomy intentionally with a rather generic and in part negatively defined notion of autonomy: "We analyze autonomous action in terms of normal choosers who act (1) intentionally, (2) with understanding, and (3) without controlling influences that determine their action."[5] One of the reasons for doing so is to acknowledge and accommodate various different and competing conceptions of autonomy. Yet, while they indeed explicitly acknowledge the importance of relational accounts of autonomy in this regard, they miss to discuss them in more detail, especially when it comes to addressing their competing implications in comparison to strictly individualistic accounts, like the resulting analysis and assessment of the influence of friends and family on a patient's decision-making process.[6] Moreover, given that autonomy essentially refers to the idea and capacity of *self-determination* and, thus, includes an *authenticity criterion* to ensure that it is actually the agent him- or herself determining his or her decisions and actions, this raises the question of how to spell out this authenticity criterion and, consequently, the notion of the agent's self in more or less relational or individualistic terms as well.[7]

Accordingly, one of the driving motivations behind putting together this volume is to address and discuss critically the implications of theories of the self and of personal autonomy for medical ethics and practice, especially for the principle of respect for autonomy, and vice versa. Certain hard cases in medical ethics prove to be exceptionally well suited to accomplish this, not only because of their challenging circumstances in general, but also because they specifically bring to the forefront contested implications of the respective underlying account of autonomy. Given the range of new possibilities in assisted reproductive technologies, such questions already have to be raised at the very beginning of life, i.e. what impact these technologies might have on the constitution of the future person's authenticity and autonomy. On the other hand, in cases of palliative care and end-of-life decisions, e.g. physician-assisted suicide, patients usually have only little control left over how they live their remaining life. Their autonomy, especially their capacity for carrying out their autonomous decisions, including bringing about their own death, may be drastically diminished or may even no longer exist at all. Moreover, finding out whether, or ensuring that, a patient's decision is actually autonomous poses major

---

[4]Beauchamp and Childress (2013). For helpful introductions to medical ethics more generally, see Schramme (2002), Schöne-Seifert (2007), Have and Gordijn (2013), Wiesemann and Simon (2013), Sturma and Heinrichs (2015), and Hope and Dunn (2018).

[5]Beauchamp and Childress (2013), 104.

[6]Cf. Beauchamp and Childress (2013), 106. It should be noted that Beauchamp and Childress address such issues partially in the context of their three other principles, i.e. nonmaleficence, beneficence, and justice. However, this already characterizes these issues as external to a person's autonomy, while relational accounts of autonomy typically aim at depicting social relations as internal characteristics of autonomy. Hence, Beauchamp and Childress essentially still presuppose an individualistic conception of autonomy.

[7]See in this regard, for example, the contributions in Kühler and Jelinek (2013a) and, for an overview, the volume's introduction, Kühler and Jelinek (2013b).

epistemic, normative, and practical problems. What (epistemic) conditions should be fulfilled in order to conclude that it is really the patient's autonomous and authentic will to die? What (normative) reasons are or should be considered decisive when it comes to determining what may be regarded as sufficiently autonomous in such cases in the first place, i.e. why should some preference or decision be regarded as autonomous and be respected? How can respect for and support of autonomy be ensured in medical practice in general if the patient is no longer able or drastically impaired in his or her capacity to utter what he or she wants, e.g. in certain cases of palliative care or in case of a comatose patient?[8]

At this point, underlying theories of the self and of personal autonomy are obviously of crucial importance because they have a fundamental impact on what kind of answers to such epistemic, normative, and practical questions may be regarded as promising or plausible. For example, could a narrative approach to the constitution of the self and to personal autonomy give decisive (epistemic and/or normative) reasons to interpret and assess as (in-)authentic and (non-)autonomous what an unconscious, comatose, or otherwise incompetent patient would want, depending on how well it fits the patient's constitutive story of his or her self? A story might, for example, explicitly include a certain personal evaluative outlook on 'death,' which could guide the (narrative) assessment of the patient's intention to die. Alternatively, based on relational accounts of personal autonomy and the self, especially friends and family seem to be able to play a crucial role in determining a patient's autonomous will, as they are already considered being partially constitutive for the patient's autonomy in general. In contrast, individualistic accounts usually dismiss external social influences when it comes to the constitution of autonomy and the self. At most, such influences are regarded as posing a threat to autonomy. Consequently, supporting and respecting a patient's autonomous preferences would have to play out quite differently in practice.

In any case, given that our personality and preferences are likely to change over time, at least to some degree, what about the normative standing and binding force of patients' advance directives to ensure long-term respect for their autonomy? On the one hand, advance directives are explicitly intended to secure respect for autonomy even if a patient is no longer able or severely impaired to make autonomous decisions. Referring to the patient's advance directive apparently provides an easy and decisive way of determining what the patient autonomously wants in the situation at hand. However, it is still a matter of debate what epistemic and normative bearing past decisions should have on current situations, especially in cases of conflict. Why

[8] As one reviewer of this volume helpfully remarked, it should be noted that some hard cases nowadays pose a flip sided challenge. In these cases, the patient explicitly wants to continue treatment, e.g. due to the patient's advance directive including that "everything be done," while physicians and other caregivers would prefer to cease treatment due to it being clearly futile. This indeed raises the familiar issue discussed in this volume, i.e. whether the patient's wish that "everything be done" may be considered sufficiently autonomous. However, given the assumed futility of further treatment, the more pressing ethical questions presumably concern matters of justice in access to specific and likely expensive health care while assuming that health care resources are scarce and need to be distributed effectively, efficiently, and fairly—questions that go beyond the scope of this volume.

exactly should an advance directive have priority over a patient's current, conflicting preference, or vice versa, e.g. in cases of dementia and major changes in the patient's personality? Moreover, substantial changes in personality, e.g. again in cases of severe dementia, raise critical questions about whether these patients can even still plausibly be held responsible for their past actions. This poses a major challenge in criminal law and for the penal system when dealing with inmates suffering from dementia. And even if such inmates are still considered responsible for their past crimes, how should their special needs for appropriate treatment be acknowledged best, especially when assuming that prisons are usually not well-equipped to handle such cases? In general, how should we, as individuals as well as society, address such challenges? Despite the complexity of these issues as a whole, keeping in mind the implications of competing theories of autonomy and the self for medical ethics and law-making at least seems to be a promising starting point.

However, respect for autonomy is not the only principle to be adhered to. For the purpose of this volume, in particular beneficence and nonmaleficence play an important role as well, again especially in hard cases such as palliative care and end-of-life decision-making. In palliative care, attempts at reducing a patient's pain and suffering through suitable medical drugs often include the unwanted side effect of diminishing the patient's capacity for autonomy (even more) or of diminishing his or her remaining life expectancy, with the latter blurring the line between treatment and physician assisted suicide in case the patient wants to die.[9]

In addition to such cases located literally at the end of life, we also often talk figuratively about someone's life having come to an end, namely in cases in which a person has not only reached a low point in life, but in which there is, apparently, also no longer any, or only very little, hope that his or her well-being will improve again.[10] This holds, for example, for people suffering from certain permanent diseases or poor health conditions, as is often the case when living—or rather only surviving—at the margins of society, like homeless people. Assuming that patients in such dire circumstances are especially vulnerable,[11] it appears to be an extra challenge when it comes to respecting their autonomy and trying to improve their well-being. Put more generally, how should we analyze the relation between autonomy and vulnerability, and what ethical consequences should be drawn from this analysis?[12] Ultimately, such cases pose not only major challenges to medical ethics in particular, but to ethics, politics, and law-making in general.

[9]In this regard, the doctrine of double effect may also play an important role. See McIntyre (2014).

[10]Article 8, "Respect for human vulnerability and personal integrity," of the *Universal Declaration on Bioethics and Human Rights* pays special attention to such individuals and groups: "In applying and advancing scientific knowledge, medical practice and associated technologies, human vulnerability should be taken into account. Individuals and groups of special vulnerability should be protected and the personal integrity of such individuals respected" (UNESCO 2005, 77). Still, no specific guidelines are mentioned that define the manner in which such persons have to be protected, especially when it comes to the relation between vulnerability and autonomy.

[11]Another example of such vulnerable positions would be embryos created in IVF (in vitro fertilization).

[12]For a discussion of some practical implications of such cases, see Mitrović (2015).

1.2 Contributions

The papers of this volume are grouped into two sections. In the first section, specific theories of the self and of autonomy are used as starting points in order to flesh out and discuss critically their implications for how to deal with hard cases in medical ethics. The contributions in the second section take the other way around and discuss crucial issues in medical ethics in order to make explicit and discuss critically their implications for formulating adequate positions in theoretical ethics or law. It should also be noted that a number of contributions explicitly consider specific German or Serbian points of view in terms of idiosyncratic social norms or laws, while other contributions aim at a more universal perspective when discussing their topic.[13]

1.2.1 Section 1: From Theories of the Self and Autonomy to Medical Ethics

In her contribution "What Is Autonomy Anyway?," Milijana Djerić takes respect for autonomy in palliative care and end-of-life decision-making usually to imply a patient's right to choose or refuse a life sustaining treatment. It is, however, close only to one sense of this notion, namely to autonomy as individual or personal liberty. This sense of autonomy is derived from Mill's work *On Liberty*, and it is important for the context of end-of-life decision-making because it provides the classical grounds for the moral obligation of a physician not to proceed without the patient's consent. It highlights the importance of the absence of external constraints and rules out the possibility of medical paternalism. However, there is another sense of autonomy, namely autonomy as rationality. This notion highlights the importance of the absence of internal constraints and refers to autonomy as decision-making capacity. It implies the absence of cognitive and psychological limitations in decision-making. Therefore, it seems that true autonomy implies that the person is free from internal as well as from external constraints. The aim of Djerić's contribution is to point out these two different notions of autonomy, to explore the role and mutual relations of these notions in the context of end-of-life decision-making, and to address the question of whether and, if so, how terminally ill patients may still be able to achieve true autonomy in this sense.

In her contribution "Personal Identity and Self-Regarding Choice in Medical Ethics," Lucie White argues that when talking about personal identity in the context of medical ethics, ethicists tend to borrow haphazardly from different philosophical notions of personal identity, or to abjure these abstract metaphysical concerns as having nothing to do with practical questions in medical ethics. In fact, as she points out, part of the moral authority for respecting a patient's self-regarding decisions can only be made sense of if we make certain assumptions that are central to a particular,

[13]For an instructive overview of different cultural perspectives on medical ethics and bioethics more generally, see Have and Gordijn (2013), section III.

psychological picture of personal identity, namely, that patients will remain psychologically connected to a certain degree with their future selves. White draws this out in detail, shows important problems with approaches in medical ethics based on alternate theories of personal identity that do not recognize this, and explores some important implications.

In his contribution "Personal Identity and the Validity of Advance Directives," Karsten Witt addresses the fundamental role of personal identity within medical ethics and especially its role and importance for certain kinds of future-directed medical decisions, namely decisions enshrined in oral or written advance directives. While it is not always clear in what sense the term *personal identity* is used in respective debates, part of the discussion is about our numerical identity through time and, at first sight, its importance seems indisputable. When we declare that a certain patient should under such-and-such future circumstances receive such-and-such medical treatment or, more frequently, that she should not receive a given treatment, whether we are that patient appears to make a crucial difference. According to conventional bioethical wisdom, medical decision-making for others is much more problematic, and in need of a different kind of justification, than deciding for ourselves. And whether the decision is self-regarding or other-regarding seems to turn on personal identity in the strict philosophical sense. While this intuitively plausible claim about personal identity's importance is widely shared, it has recently been criticized by various bioethicists for its seemingly troubling consequences for advance directives when combined with a psychological-continuity account of personal identity: the signer of an advance directive might not be numerically identical with her severely demented or permanently comatose 'biological successor', possibly depriving the directive of its moral authority. Still, Witt defends the initial claim and mainly argues that many objections against it are variations of the idea that what is crucial for the validity of advance directives is not whether signer and patient *are* numerically identical, but whether certain people in the patient's environment *believe* that they are. Yet, this idea is problematically vague, Witt argues, as it borders on incoherence, and has troubling ethical consequences. Critics of the initial claim, therefore, have two options left: they might save part of their critique by embracing conventionalism about personal identity or they might claim that advance directives are not about self-determination but can still play an important part in decisions about the patient's best interests.

In his contribution "Narratives in Flux. Why Patients' Life Stories Do Not Provide Decisive Instructions in Cases of Surrogate Decision-Making," Michael Kühler critically discusses the role narrative accounts of the self might play in surrogate decision-making. Medical cases in which patients are not or no longer able to express their interests and also have not given a prior account of them present us with the question of how to respect their autonomy, i.e. how the patient would probably want to be treated, were he or she able to decide for him- or herself. In such situations it seems natural to refer to a notion of the person's identity or self, for based on this we might extrapolate how the person would decide in an authentic manner. In this regard, narrative theories of the self seem to be able to provide a reasonable answer. Knowing a person's life story seems to suggest a simple and convincing way to determine her authentic interests based on how her story, i.e. her narrative identity, would continue in a plausible and coherent way. Kühler, however, argues that narrative accounts of the self inherently fail to provide a simple and convincing answer in such cases. After distinguishing between mere expressive narrative accounts and constitutive narrative accounts, he argues that expressive narrative accounts do not present an independent analysis of the constitution of the self to begin with and are at best a helpful epistemic tool to find out what the patient's autonomous preference likely would be. Yet, even this only holds under the contestable assumption that the patient's self remains stable. Constitutive narrative accounts, on the other hand, while indeed presenting an independent account of the constitution of the self, either revert back to an expressive narrative account if the idea of one's free authorship of one's life story is rejected, or they fail to provide any reason for why one should choose a certain version of the patient's life story as the newly to be constructed one. Instead, Kühler concludes that a person's life story remains in a constant state of flux, which is why also constitutive narrative accounts are unable to provide surrogate decision makers with a well-justified answer which treatment option to choose in order to respect the patient's autonomy based on who he or she is.

In her contribution "The ART of Authenticity," Birgit Beck starts off with the fact that recent progress in biomedical research in the fields of stem cell technology and genome editing has entailed a revival of ethical concerns regarding the autonomy and authenticity of future persons who might be created by prospective novel means of assisted reproductive technologies (ART) on the basis of, e.g. in vitro gametogenesis or germline interventions. In this regard, critical authors refer to the Habermasean concern that persons who are "made" by technological means instead of coming into existence in a "natural" way could be deprived of forming a self-concept of being autonomous and authentic "authors of their life histories". The notion of being the "author" of one's life history appears to relate to the idea that we can—somehow and within certain limits—actively influence who we are. It is hard to see why this idea could not include the possibility of integrating the circumstances of one's coming into existence—be they "artificial" or "natural"—in an authenticity-preserving way. Given that (future) persons are aware of those circumstances, Beck argues that they should be able to adapt their life history accordingly. Assuming a relational structure of autonomy and authenticity in general, it appears plausible that no person's life narrative and self-concept are constituted and maintained in an isolated and individualistic manner. Quite the contrary, Beck argues, (self-)ascriptions of autonomy and authenticity require adequate social relationships in the first place. From this assumption, it follows that, should (future) persons conceived by means of novel ART be deprived of a confident self-concept and feel "objectified" by their "creators", this would be due to prejudice rather than actual lack of essentialist authenticity. For Beck, the Habermasean concern can thus be regarded as revealing a structural social problem rather than one raised by biotechnology per se, and it can be smoothed out by Habermas' own conceptual conditions for the development of personal identity, authenticity, and autonomy.

In the final contribution to the first section, "Remainders of the Self: Consciousness as a Problem for Neuroethics," Marco Stier firstly notes that neuroscience has made great progress in recent years in detecting mental processes using electrophysiological and imaging techniques. One of the major breakthroughs in this regard has been the detection of residual consciousness in persons diagnosed with a vegetative state. However, Stier argues that this very success gives rise to a number of theoretically as well as ethically crucial follow-up questions. What neuroimaging can provide are highly artificial, mathematically processed images and not direct information about the patient's mind. Therefore, it is far from clear whether and to what degree behaviorally unresponsive patients are indeed conscious. This is all the more so since we are confronted with a double mediation: one between the person's mind and the image of the brain and the second between the image and something that we interpret as a meaningful answer of the patient to a question of the physician. For the patient in question it is of vital importance that his ability for consciousness is neither over-nor underrated since both can lead to significant suffering. Above that, residual consciousness or remainders of the person's self—even if reduced and fragmented—have to be taken into account for treatment or end of life decisions. Unfortunately, as Stier points out, there is no consensus in neuroscience about what "consciousness" and "being conscious" really means. To be sure, there is a long tradition regarding notions like these in philosophy. However, philosophical insight does not find its way into neuroscience easily. Against this background, Stier's contribution aims at connecting philosophical and neuroscientific understandings of consciousness in order to better understand what may go on inside an otherwise unresponsive patient. The notion of a "self" that might be preserved even if the person in question is not fully conscious or only has some basic form of consciousness will serve as a working concept for this problem. As Stier emphasizes, neuroethics has to solve the conceptual problem of consciousness in order to be able to solve the clinical problem of consciousness and with it the ethical problem of autonomy and the self.

1.2.2 Section 2: From Medical Ethics to Theories of the Self and Autonomy

In the first contribution to section 2, "Ethical Issues Concerning Patient Autonomy in Clinical Practice," Alfred Simon takes up the perspective of clinical practice and discusses the importance of respect for patients' autonomy from this angle. He starts by mentioning that, until a few decades ago, physicians and nurses involved in ethical decision-making primarily considered the principles of beneficence and nonmaleficence. Respect for the patient's autonomy clearly played a subordinate role. In traditional ethical codes such as the Hippocratic Oath or the Declaration of Geneva of the World Medical Association of 1948, no reference is made to the patient's wishes. This has changed radically: In line with the shift in values towards greater individuality and personal responsibility observed in the Western world in the second half of the twentieth century, patient self-determination became more and more important. Legal decisions and modern medical ethics emphasize that patient autonomy overrides what physicians and nurses consider best for the well-being of the patient. Informed consent is now a widely acknowledged normative standard in medical ethics, and patient autonomy finds its concrete expression in clinical practice in this concept of informed consent. Simon introduces the concept and discusses central issues concerning patient autonomy in clinical practice: What can health care professionals do to encourage patient autonomy? What are possible limits to patient autonomy? How can a patient's capacity to consent be assessed? How should health care professionals deal with patients who have temporarily or completely lost their capacity to consent? What is the importance of advance directives? What are possible problems concerning advance directives in clinical practice, and how can they be solved?

By comparing articles from Serbian legislation, Hajrija Mujovic analyzes in her contribution "Patients' Rights to Refuse Medical Treatment Through the Act of Advance Directives" the (im)possibility of patients' rights to refuse medical treatment through the act of advance directives. Serbia's legislation on patients' rights mostly adheres to contemporary standards. But the practice of implementation shows some misunderstandings when it comes to the question of whether the rights stipulated in the Patients' Right Act (2013) completely apply to the persons in need of palliative care or in end-of-life care. Besides, there are issues particularly important for terminally ill patients, including the form of possible advance directives (Article 16) and passive euthanasia (Article 28), which still need to be discussed and revised. In fact, Serbian legislation treats so-called "active euthanasia" simply as "euthanasia" in general. Therefore, passive euthanasia, which implies withholding life-saving treatment from a patient at the patient's will, does not count as euthanasia. Following Serbian law, it will be shown that this creates a problem in terms of the patient's right to refuse medical treatment. A dying patient also exercises basic rights and human dignity in all these cases, which are to be respected as well. One of these rights is the right to provide an advance directive, which should be recommended to be revised and fully accepted in Serbian law.

Taking up the issue of advance directives, Oliver Hallich, in his contribution "Prolonged Autonomy? The Principle of Precedent Autonomy and the Binding Force of Advance Directives in Dementia," takes as starting point the fact that dementia has become a central problem of health care in modern societies. Currently, 47.5 million people worldwide suffer from some or other form of dementia, and it is projected that 75.6 million people will have it in 2030. Against this background, the question of the binding force of advance directives in dementia will be of growing importance in future years. Defenders of the view that dementia advance directives should be seen as binding often invoke the principle of precedent autonomy to support their view. According to this principle, it is the right of a competent individual to make decisions for a later time once competence has been lost. However, Hallich criticizes this principle of precedent autonomy and argues that it cannot be upheld because there are some cases in which it is morally legitimate to act against the will of the competent patient that is expressed in an advance directive pertaining to a later life stage in which competence has been lost. First, he paves the way for a criticism of the principle of precedent autonomy by laying bare its fundamental premise, namely the "dominance thesis," according to which critical interests dominate experiential interests in cases of conflict between these two kinds of interests. He then distinguishes between various ways in which critical and experiential interests may relate to each other and argues that the dominance thesis holds true only with regard to some but not all of these constellations. In focusing on a specific scenario with regard to which the dominance thesis turns out to be false, he argues that in some cases of a conflict between critical and experiential interests the latter ones negate the normative authority of the former. Based on a rebuttal of two objections to this view, he concludes that we should answer the normative question of whether advance directives in dementia should be seen as binding with a qualified "no."

In their contribution "Individual and Collective Decision-Making in Palliative and End-of-Life Care," Zoran Todorović and Dragana Protić start with the assumption that the distinction between individual and collective decision-making strongly influences the bioethical foundations and perspectives of palliative and end-of-life care. Kemp and Rendtorff[14] promoted an integrated approach to basic ethical principles and discussed that autonomy implies the capacity to make your own decisions about your own life. However, such a decision depends on the interaction or collaboration with others, including their values. In addition, autonomy is interconnected with vulnerability, and individuals are not always able to judge the treatment process and understand and weigh different perspectives and possibilities. Against this background, Todorović and Protić explore the corresponding interplay between individual and collective decision-making in palliative and end-of-life care from the point of view of medical practice.

In her contribution "Who Should Take Care of Offenders with Dementia: Some Thoughts on Fading Selves and the Challenge of Responsibility Interpretations," Annette Dufner investigates the way in which our understanding of a dementia patient's self holds relevance to issues of punishment and responsibility. This topic is motivated by the fact that some countries with particularly large prison populations—such as the United States—are starting to build specialized prison tracts for inmates with dementia. In other countries that do not have such specialized facilities, authorities are trying to find the least badly equipped facility for such patients, and they are turning to ordinary retirement homes, forensic hospitals as well as ordinary psychiatric and geriatric hospital wards. The problem is expected to become increasingly urgent as the population ages and the number of dementia patients increases. Dufner analyzes the way in which justifications of legal (or private) punishment for offenders with dementia can depend on an account of relevant psychological features of the self. As she argues, especially retributivist and expressivist justifications of punishment require the offender's ability to comprehend that he or she is being punished for a particular action in the past, and that it was him- or herself who committed this action. In the second part of the paper, Dufner distinguishes between different accounts of responsibility and argues that accounts of relevant features of the self are also needed here to answer the question of whether offenders with dementia are still responsible for past or current inappropriate behavior. In the final part of the paper, Dufner argues that certain puzzles of responsibility interpretation as well

[14] See Rendtorff (2002) and Kemp and Rendtorff (2008).

as potentially existing private punishment intuitions among caretakers can make it plausible to relieve certain caretakers from primary responsibility for offenders with dementia, especially caretakers who belong to a relevant former victim group of the offender.

In the final contribution to the volume, "Understanding 'Euthanasia' Across Various Medical Practices," Veselin Mitrović takes as starting point that end-of-life decisions and assisted suicide are often equated with 'euthanasia.' In everyday parlance of social actors, the term euthanasia is understood rather broadly, even lumped together with other medical procedures. Still, Mitrović argues that 'intended merciful death,' whether we like the definition or not, ought not to be equated with other practices. Although all of these medical procedures result in the destruction of potential or actual life, the reasons behind such actions could be quite different from empathy or mercy, making the acceptance and advocacy of a problematic definition and understanding of euthanasia the subject of ethical and social debates and analyses. When considering the Universal Declaration on Bioethics and Human Rights (from the 33rd session of the General Conference of UNESCO, (2005)),[15] specifically the principle of human dignity and rights (article 3), the principle of benefit and harm (article 4) as well as the principle of autonomy (article 5), the debate stretches out to include also vulnerable groups in general, which in the contemporary context range from homeless persons and other marginalized groups to embryos created during IVF (in vitro fertilization). In his article, Mitrović presents two case studies, chosen from ten personal stories of former and current IVF procedure patients. In all ten narratives, interlocutors equate abortion with embryo reduction, and both of those with euthanasia. Mitrović analyzes their perspective to embryos that were not implanted, as well as similarities and differences in their views regarding the activities in the cases of implanted embryos (twin and triplet pregnancies).

References

Beauchamp, Tom L., and James F. Childress. 2013. *Principles of Biomedical Ethics*, 7th ed. Oxford: Oxford University Press.

Buss, Sarah. 2013. Personal Autonomy. In *The Stanford Encyclopedia of Philosophy, Summer 2013*, ed. Edward N. Zalta. https://plato.stanford.edu/archives/sum2013/entries/personal-autonomy/.

Christman, John, and Joel Anderson (eds.). 2005. *Autonomy and the Challenges to Liberalism*. Cambridge: Cambridge University Press.

Christman, John. 2015. Autonomy in Moral and Political Philosophy. In *The Stanford Encyclopedia of Philosophy, Spring 2015*, ed. Edward N. Zalta. https://plato.stanford.edu/archives/spr2015/entries/autonomy-moral/.

Eyal, Nir. 2012. Informed Consent. In *The Stanford Encyclopedia of Philosophy, Fall 2012*, ed. Edward N. Zalta. https://plato.stanford.edu/archives/fall2012/entries/informed-consent/.

Have, A.M.J. Henk ten, and Bert Gordijn (eds.). 2013. *Handbook of Global Bioethics*. New York: Springer.

[15] See UNESCO (2005).

Hope, Tony and Dunn, Michael. 2018. *Medical Ethics. A Very Short Introduction*, 2nd ed. Oxford: Oxford University Press.
Kemp, Peter, and Jacob Dahl Rendtorff. 2008. The Barcelona Declaration. Towards an Integrated Approach to Basic Ethical Principles. *Synthesis Philosophica* 23: 239–251.
Kühler, Michael and Jelinek, Nadja. (eds.). 2013a. *Autonomy and the Self, Philosophical Studies Series 118*. Dordrecht: Springer.
Kühler, Michael and Jelinek, Nadja. 2013b. Introduction. In *Autonomy and the Self*, ed. Michael Kühler and Nadja Jelinek, ix–xxxvi. Dordrecht: Springer.
Mackenzie, Catriona, and Natalie Stoljar (eds.). 2000. *Relational Autonomy: Feminist Perspectives on Automony, Agency, and the Social Self*. Oxford: Oxford University Press.
Maclean, Alasdair. 2009. *Autonomy, Informed Consent and Medical Law: A Relational Challenge*. Cambridge University Press.
McIntyre, Alison. 2014. Doctrine of Double Effect. In *The Stanford Encyclopedia of Philosophy, Winter 2014*. ed. Edward N. Zalta. https://plato.stanford.edu/archives/win2014/entries/double-effect/.
Mitrović, Veselin. 2015. Resilience: Detecting Vulnerabilities in Marginal Groups. *Disaster Prevention and Management* 24: 185–200.
Rendtorff, Jacob Dahl. 2002. Basic Ethical Principles in European Bioethics and Biolaw: Autonomy, Dignity, Integrity and Vulnerability—Towards a Foundation of Bioethics and Biolaw. *Medicine, Health Care and Philosophy* 5: 235–244.
Schöne-Seifert, Bettina. 2007. *Grundlagen der Medizinethik*. Stuttgart: Alfred Kröner.
Schramme, Thomas. 2002. *Bioethik*. Frankfurt am Main: Campus.
Stoljar, Natalie. 2015. Feminist Perspectives on Autonomy. In: *The Stanford Encyclopedia of Philosophy, Fall 2015*. ed. Edward N. Zalta. https://plato.stanford.edu/archives/fall2015/entries/feminism-autonomy/.
Sturma, Dieter, and Bert Heinrichs (eds.). 2015. *Handbuch Bioethik*. Stuttgart: Metzler.
Taylor, James Stacey. (ed.). 2005. *Personal Autonomy. New Essays on Personal Autonomy and Its Role in Contemporary Moral Philosophy*. Cambridge: Cambridge University Press.
UNESCO. 2005. Universal Declaration on Bioethics and Human Rights. https://unesdoc.unesco.org/images/0014/001428/142825e.pdf.
Wiesemann, Claudia, and Alfred Simon (eds.). 2013. *Patientenautonomie: Theoretische Grundlagen—Praktische Anwendungen*. Münster: Mentis.

Michael Kühler is Assistant Professor at the University of Twente, Netherlands, and "Privatdozent" (roughly equaling Associate Professor) at Münster University, Germany. His research interests include ethics, medical ethics, metaethics, and political philosophy. His recent publications include "Technological Moral Luck," in: Beck, Birgit/Kühler, Michael (eds.): *Technology, Anthropology, and Dimensions of Responsibility*, Techno:Phil 1, Stuttgart: Metzler, 2020, 115–132, URL: https://doi.org/10.1007/978-3-476-04896-7_9, "Romantische Liebe und die Freiheit zu gehen," in: Dietz, Simone/Foth, Hannes/Wiertz, Svenja (eds.): *Die Freiheit zu gehen. Ausstiegsoptionen in politischen, sozialen und existenziellen Kontexten*, Wiesbaden: Springer, 2019, 195–222, "*Modus Vivendi* and Toleration," in: Horton, John/Westphal, Manon/Willems, Ulrich (eds.): *The Political Theory of Modus Vivendi*, Dordrecht: Springer, 2019, 235–253, "Toleranz und/oder Paternalismus im engeren sozialen Nahbereich?," in: *Zeitschrift für Praktische Philosophie* 4, 2017, 63–86; and "What if I Cannot Do What I Have to Do? Notions of Personal Practical Necessity and the Principle 'Ought Implies Can'," in: Bauer, Katharina/Mieth, Corinna/Varga, Somogy (eds.): *Dimensions of Practical Necessity. "Here I Stand. I Can Do No Other,"* Houndmills: Palgrave Macmillan, 2017, 87–107.

Veselin Mitrović is a senior research associate at the Institute of Social Sciences at the University of Belgrade. His research activities feature topics in Bioethics, Disaster Study, Sociology of Culture and Social Action in different strata (from marginal to elite). His recent publications include "Ethics and Floods: A Systematic Review." *Disaster Medicine and Public Health Preparedness.* 13/4, 2019, 817–828. "Parent's Religious and Secular Perspectives on IVF Planning in Serbia," *Journal for the Study of Religions and Ideologies* 15/43, Spring 2016, 48–81; "From Global Bioethics to Neuroethics," *TEME* XL/4, 2016, 1457–1475 (Serbian); "Resilience: detecting vulnerability in marginal groups," *Disaster Prevention and Management* 24/2, 2015, 185–200; "The Contingency of the 'Enhancement' Arguments: The Possible Transition From Ethical Debate to Social and Political Programs," *Journal for the Study of Religions and Ideologies* 13/37, Spring 2014, 93–124; "Ethos of the Elite and Future of Serbia," *Balkanske sinteze* 1/1, 2014, 167–189 (Serbian); "Nanotechnologies and Global Survival," *Sociologija* LIV/1, 2012, 37–54 (Serbian); "The Myth of Moral Enhancement: Back to the Future?," *Filozofija i društvo* XXIII/2, 2012, 111–123 (Serbian); "Human Enhancement: Toward the Creation of Patterns of Injustice?" Zeller, Jörg/Riis, Ole Preben/Nykänen, Hannes (eds.): *Issues in Theoretical and Applied Ethics*, Aalborg: Aalborg University Press, 2013; "Study," *Sociology in Serbia 1959–2009, Institutional Development*, 59–107, (ed.) Marija Bogdanović, (Beograd: Službeni glasnik i Filozofski fakultet Univerzitet u Beogradu, 2009) (Serbian). He is also author of the following books written in Serbian: *Apathetic Society* (2015); *The Stride of Bioethics, New Bio-Technologies and Social Aspects of the "Enhancement" of the Healthy* (2012) and *Jazz as Socio-Cultural Improvisation—A Qualitative Research of Social Mobility* (2012).

Part I
From Theories of the Self and Autonomy to Medical Ethics

Chapter 2
What is Autonomy Anyway?

Milijana Đerić

2.1 Introduction

Autonomy is an important concept for numerous disciplines.[1] Besides philosophy, it occupies an important place in sociology, psychology, biology, computer sciences, etc. Within philosophy, the concept of autonomy is central to a range of philosophical areas such as metaphysics, ethics, social and political philosophy. And within ethics, questions regarding the nature and value of autonomy have important consequences for questions of moral responsibility and free will, in addition to a variety of concerns in practical ethics. This is why there is no unique conception of autonomy, but rather a variety of conceptions, shaped around on the concerns and questions in the field in which it is utilized.

In this paper, I will be examining the concept of autonomy in the context of medical ethics, in particular, pertaining to end-of-life decision-making and I will be advocating individualistic conception of autonomy, First, I examine the concept of autonomy and shed light on the etymology of the concept as well as on its diverse uses in contemporary literature. I highlight the distinction between Kant's conception of autonomy as "moral autonomy" and Mill's conception of autonomy as "personal autonomy". Then, I show that the concept of autonomy invoked in medical ethics is the individualistic concept, based primarily on Mill's liberty principle. My goal is to

[1] I thank the editors of this volume, Michael Kühler and Veselin Mitrovic, and Lucie White for their helpful comments and suggestions on the first version of this paper.

This paper is funded by The Ministry of Education, Science and Technological Development of the Republic of Serbia, as a part of the project "Dynamic Systems in Nature and Society: Philosophical and Empirical Aspects", no. 179041.

M. Đerić (✉)
Institute for Philosophy, Faculty of Philosophy, University of Belgrade, Belgrade, Serbia
e-mail: milijana.djeric@f.bg.ac.rs

M. Kühler and V. L. Mitrović (eds.), *Theories of the Self and Autonomy in Medical Ethics*, The International Library of Bioethics 83,
https://doi.org/10.1007/978-3-030-56703-3_2

show that a patient's right to self-determination, secured through informed consent, is not based upon an adequate conception of autonomy in medical ethics. An adequate individualistic conception of autonomy in medical ethics and in particular in end-of-life-decision-making must be two-dimensional. It must recognize the distinction between agent-autonomy and action-autonomy and highlight the importance of the absence of internal constraints as well as external constraints. After reviewing two objections against invoking action-autonomy as a means to understanding end-of-life decision-making, I reject a "perfect autonomy" standard for action-autonomy and introduce a "sufficient autonomy" standard in this context. The advantage of this model over the previous one lies in its adaptability in a variety of medical circumstances and background conditions. According to this model, in order to be autonomous, the terminally ill patient's choice to die does not need to be perfectly autonomous; it can be simply sufficiently autonomous.

2.2 An Overview of Conceptions of Autonomy

Although autonomy has a prominent role in medical ethics (in particular, in debates on informed consent and medical paternalism) the original concept of "autonomy" had nothing to do with health care or indeed with individuals. The term is derived from the Greek words *autós* and *nomos*, meaning literally, "self-rule". It was used to describe the self-sufficiency and self-government of Greek polis or city-state and its political capacity of making its own laws and maintaining political and economic independence. For example, Aristotle held that self-sufficiency or *autarkēs* is independence from external forces and influences (e.g. need, utility, etc.): "[…] we take what is self-sufficient to be that which on its own makes life worthy of choice and lacking in nothing" (2000, 1097b), and it is primary good and final end of a city-state: "Again the end, or final cause, is the best and self-sufficiency is both the end, and the best" (1998, 1252b27). According to Aristotle, the city-state achieved the highest degree of such independence and provides all that is necessary for a good life, i.e. a life in accordance with virtues.

In contemporary literature, the concept of autonomy is a vague and multifarious concept, used in many different ways and with many different meanings, as acknowledged both by autonomy theorists (Dworkin 1988, 2015; Feinberg 1986, Taylor 2009) and bioethicists (Engelhardt 2001; Gracia 2012). Therefore, an overview of its numerous distinct meanings will be a good starting point for our reflection on its meaning in medical ethics. Gerald Dworkin holds that the concept of autonomy is used:

> […] sometimes as an equivalent of liberty (positive or negative in Berlin's terminology), sometimes as equivalent to self-rule or sovereignty, sometimes as identical with freedom of the will. It is equated with dignity, integrity, individuality, independence, responsibility, and self-knowledge. It is identified with qualities of self-assertion, with critical reflection, with absence of external causation, with knowledge of one's own interests (1988, 6).

Dworkin also points out that the only common feature of autonomy shared by these distinct concepts is that it is a feature of a person and a desirable quality to have (1988, 6). However, it is important to note that this assertion is problematic because autonomy also can be a feature not of persons but of polities (as Aristotle thought) or the will (as Kant thought). Moreover, as we will see, it can also be a characteristic of a person's acts or choices. Although Dworkin is correct in saying that autonomy is often understood to be equivalent to self-rule or sovereignty, this should be taken simply as "an etymological point rather than a philosophical one" (Taylor 2009, 21). Joel Feinberg also noticed that the taxonomy of conceptions of autonomy is extensive and diverse. According to him, it includes: self-possession, distinct self-identity (individuality), authenticity, self-creation, self-legislation, moral authenticity, moral independence, integrity, self-control, self-reliance, self-generation, responsibility for self (1986, 31–44). However, his own account of autonomy includes the analysis of only four different meanings. He distinguishes between autonomy as the *capacity* to govern oneself, autonomy as the *actual condition* of self-government, autonomy as an *ideal of character*, and autonomy as the *right to self-determination* (Feinberg 1986, 27–51). Although they differ about in what this core concept consists of, most authors agree that the core of Feinberg's account of autonomy is autonomy as an *actual condition.* Rather than being a separate and different concept, the four meanings are closely related in sense that autonomy as an *actual condition* is the one upon which three others are based on (May 1994, 137–138; Taylor 2009, 20–21). A further theorist, Diego Gracia, outlines four distinct conceptions of autonomy: *political autonomy*, *metaphysical autonomy*, *legal autonomy* or informed consent, and *moral autonomy* (2012, 58–63) and bioethicist Tristram Engelhardt details three conceptions: autonomy as a *good*, as a *right-making condition* and as element of *human flourishing* or virtue (2001, 286–291).

Most recently, the conceptualization of autonomy has emerged in relational terms. This has resulted in the substantial and mighty debate by philosophers who criticized an individualistic understanding of autonomy and called for its reconceptualization. Feminists who claimed that the individualistic concept of autonomy must be reconfigured so as to be more sensitive to relations of care, interdependence and mutual support, have been especially vocal in this (Mackenzie and Stoljar 2000; Meyers 2002; Oshana 2016). There is a vast and different usage of the concept of autonomy within relational theories of autonomy, but the main distinction is the one between procedural and substantive approaches to autonomy. Procedural conceptions have been prominent in standard literature on autonomy from the 1980s (Dworkin 1988; Frankfurt 1988) to the present (Friedman 2003; Christman 2004, 2009). They are "content-neutral" and "non-normative" since according to their advocates a person must endorse no value or set of preferences in order to be autonomous. On the contrary, either in its strong (Benson 2005; Charles 2010) or weak form (Richardson 2001; McLeod 2002), substantive conceptions of autonomy are value-oriented. According to strong substantive conceptions, "the contents of the preferences or values that agents can form or act on autonomously are subject to direct normative constraints" (Benson 2005, 133), while weak substantive conceptions do not pose external conditions as necessary conditions of autonomy.

2.3 "Moral Autonomy" and "Personal Autonomy"

The two most prominent conceptions of autonomy are "moral autonomy" and "personal autonomy". These notions of autonomy rely on two different philosophical foundations: Kant's deontological ethics and Mill's liberalism. Regarding medical ethics, some authors were relying on both sources simultaneously (Beauchamp and Childress 1979, 56–59)[2] while others were giving a greater importance to Kant's ethics and mistakenly argued that an essential part of it is personal autonomy (Gillon 1985). However, as contemporary Kant scholars (O'Neill 2002, 83–86; Secker 1999, 43–66) and autonomy theorists (Taylor 2005, 1) emphasize, these two conceptions of autonomy—moral autonomy and personal (individual) autonomy—are quite different.[3]

"Moral autonomy" refers to Kant's categorical imperative, and is expressed in any action based upon a principle that could be universally accepted by all rational beings. His principle of autonomy is thus: "Not to choose otherwise than so that the maxims of one's choice are at the same time comprehended with it in the same volition as universal law" (Kant 2002, 58). According to Kant's conception, autonomy is a characteristic of the principle behind the action rather than of action itself and is not characteristic of agents at all.[4] Contemporary Kant scholars emphasize that Kant never spoke of *autonomous persons* or *autonomous individuals*, but rather of the *autonomy of principles* and the *autonomy of the will* (Hill 1992, 79–96; Hill 2013, 15–31; O'Neill 2002:83).[5] In his *Groundwork for the Metaphysics of Morals*, Kant explains that autonomy is a property of the will of all rational beings. According to him, autonomy of the will is the sole principle of morality and of all duties which conform to it. The will, in order to be autonomous, must be motivated by reason alone. Thus, it should not be influenced by external factors in determining duty: "Autonomy of the will is the property of the will through which it is a law to itself (independently of all properties of the objects of volition)" (Kant 2002, 58). However, if the will is heteronomous, i.e. influenced by external considerations, the moral value of the action and determination of the duty becomes contingent. In such a case, the will

---

[2]Beauchamp and Childress explicitly invoke both Kant's deontology and Mill's liberalism in their formulation of the principle of autonomy in the first edition of *Principles of Biomedical Ethics*, but in subsequent editions, the Kantian view of autonomy became less important than the Millian view.

[3]For a discussion of other various contemporary conceptions of autonomy and the ways in which they differ from Kant's conception see Hill (2013, 15–32) and Schneewind (2013, 146–169).

[4]Nevertheless, one might point out that Kant emphasizes the autonomy of persons in speaking of our capacity to act autonomously. According to him, all finite rational beings have the capacity to give themselves a law for action, which can only be the moral law. This certainly speaks in favor of Kant's conception of autonomy as characteristic of agents, i.e. personal autonomy. However, Kant speaks of autonomous persons in the sense only of moral, not personal autonomy. For him, persons have the capacity to act autonomously if and only if they have the capacity to act morally. Therefore, autonomous persons are only those able to take the categorical imperative as their fundamental law.

[5]This is the reason why Onora O'Neill uses the term *principled autonomy* when referring to Kant's conception of autonomy (O'Neill 2002, 85). Some other references to Kant's conception of autonomy include terms such as *will autonomy* (Dan-Cohen 2002, 135) and *metaphysical autonomy* (Gracia 2012, 58–59).

cannot be the basis of the duty and it is opposed to the principle of morality (Kant 2002, 58–59).

Therefore, Kant's test of universalizability is a test of self-sufficiency of the principle of action (Kant's maxim). If the principle can be universalized, it is independent of external factors, and based purely on reason. However, if it cannot be universalized, the moral value of the principle is dependent on the contingent circumstances. Such a principle is not autonomous and agents who act upon it are not acting in accordance with autonomy.

"Personal autonomy", on the other hand, is based upon Mill's liberalism, and is expressed in his anti-paternalistic liberty principle.[6] The principle is that:

> The only purpose for which power can be rightfully exercised over any member of a civilized community, against his will, is to prevent harm to others. His own good, either physical or moral, is not a sufficient warrant. [...] The only part of the conduct of any one, for which he is amenable to society, is that which concerns others. In the part which merely concerns himself, his independence is, of right, absolute. Over himself, over his own body and mind, the individual is sovereign (Mill 2009, 18-19).

However, for Mill, the application of the principle is not absolute and limitless; the absolute right to the liberties detailed by Mill is restricted to people that are "of full age, and [possess an] ordinary amount of understanding" (Mill 2009, 127). That is, the principle is not applicable to children, the mentally ill, or those with limited cognitive capacities. Also, those who are not aware that they are harming themselves are also excluded from the absolute right to liberty concerning self-regarding action. In his famous example, Mill describes a person attempting to cross a dangerous and unsafe bridge in circumstances in which it is uncertain whether he is aware of the danger. His conclusion is that preventing the person from crossing the bridge is permissible only if he is unaware of the danger, because "liberty consists in doing what one desires, and he does not desire to fall into the river", but if he is aware of it he should be permitted to cross the bridge (2009, 163–164).

The main purpose of Mill's liberty principle is to promote personal autonomy, i.e. a person's freedom from external constraints. The voluntary choices of competent adults shouldn't be constrained by others or by the state. The difference between personal and moral autonomy is therefore vast. One key difference is the absence of a moral foundation for personal autonomy. Unlike Kant, (who was concerned with autonomy as a fundamental moral principle) Mill was concerned with autonomy needed for personal freedom and political civil liberty, not morality.

[6]Although Mill never uses the term "autonomy" in *On Liberty*, contemporary authors often use it in reviewing and explicating his position. See, for example, Donner (2008), Gray (1996) and Turner (2013).

2.4 Autonomy in Medical Ethics and the Distinction Between Agent-Autonomy and Action-Autonomy

The understanding of autonomy in medical ethics (used to evaluate issues ranging from euthanasia and abortion to the sale of organs and other body parts) derives primarily from Mill's liberty principle rather than Kant's principle of universal moral law. Autonomy is often invoked in the context of a principle of respect for autonomy, i.e. an obligation to respect the patient's individual desires and capacity for self-determination. This principle has been established and promoted in numerous declarations and reports (World Medical Association 2013; Kemp and Rendtorff 2008; The National Commission for the Protection of Human Subjects of Biomedical and Behavioral Research 1978) as well as by many bioethicists (Gillon 2003; Taylor 2009). Even though some authors assert that the principle of respect for autonomy shouldn't be taking precedence over other ethical principles (such as beneficence, non-maleficence and justice) (Beauchamp and Childress 2013) or advocate non-principle-based (Harris 2003), communitarian (Gauthier 2000, Callahan 2003a, b) or virtue-ethics-based approaches to medical ethics (Campbell 2003), today there is little doubt that the principle of respecting a patient's autonomy is regarded as "the first among equals".[7]

The principle of respecting a patient's autonomy requires that patients are allowed to act upon their own decisions, in accordance with self-chosen plans for their own lives. In practice, this means that patients have the right to make decisions about their medical treatment without their physician trying to influence their decision. Respect for the patient's individuality and right to self-determination are compatible with physicians providing information to or educating the patient concerning the various options available to them, but it does not allow physicians to make the decisions on behalf of the patient. This individualistic conception of autonomy as protecting the patient's individuality and right to self-determination is of thus a central part of medical ethics, for it establishes the necessity of obtaining informed consent from the patient, and rules out the possibility of medical paternalism (Ho 2008; Ikonomidis and Singer 1999; Krishna et al. 2015). However, it is not sufficient as a basis for dealing with all autonomy-related concerns in medical ethics. Regarding that it is not always the case that an act performed by an autonomous agent is itself autonomous, the conception of autonomy must include more than the patient is free from external constraints and other individuals. Namely, an autonomous, externally unconstrained agent can be coerced by internal influences. He can be constrained by some kind of cognitive incapacity or can be ignorant of significant features of his act. In such a case, he may not act autonomously. Physicians often meet patients who are generally autonomous, yet sometimes make non-autonomous decisions. This is caused by the fact that the threat to patient's autonomy doesn't come only by external factors—coercion, manipulation, and persuasion by others (family, physicians, etc.),[8] but also

[7]For the phrase "first among equals" used in this context see Gillon (2003).

[8]For more on nature and different types of coercion, persuasion and manipulation see: Faden and Beauchamp (1986, 337–368).

by internal factors—the very condition of the patient, his terminal disease, mental illness, etc. Namely, his disease can influence his decision-making capacity and deprive his decisions and choices of autonomy. And when it comes to the terminal stage of illness, the patient is hardly ever free from internal constraints and autonomy-undermining factors such as depression, delirium, dementia, severe pain, suffering, feelings of worthlessness, etc. As many studies have shown, these internal factors influence the decision-making capabilities of patients with severe or immediately life-threatening disease (Schaeffer et al. 1996; Hustey and Meldon 2002; Pisani et al. 2003; Raymont et al. 2004).

Therefore, an adequate conception of autonomy in medical ethics cannot be concerned exclusively with the obtaining a patient's informed consent. Instead, an adequate understanding of autonomy must encompass two components: "agent-autonomy" and "action-autonomy". This means that the concept of autonomy must be applied not just to agents, i.e. patients, but also to patient's actions, i.e. their decision or choices. More particularly, we must ask not just the question about patient's autonomy—is he free from external and internal influences to exercise his right to informed consent, but also the question about his consent's autonomy—is his decision or choice free from internal and external constraints? In what comes, I will focus mainly on action-autonomy within an end-of-life decision-making context and on internal influences. I will suppose here that the terminally-ill patient's decision or choice to die is free from external influences, and deal with the question of if it is ever free from autonomy-compromising internal influences.

2.5 The Problem of Action-Autonomy in End-of-Life Decision Making

Within the end-of-life decision making, in current consent cases or "current end of life decisions" (Colburn 2013, 77) in which there is no prior consent or advance directive from the terminally ill patient, action-autonomy is a subject to two main objections.

The first objection is that in such cases action-autonomy is never possible since the illness and pain are determining factors in a patient's decision to die. The terminally ill patient's decision to die is never autonomous because it is caused by his unbearable pain and suffering (Campbell 1999, 244) and disrupted life-plans and sense of self (Pellegrino and Thomasma 1981). The illness "deprives the patient of his distinctly human freedoms—to act, to make his own choices" (Pellegrino and Thomasma 1981, 252) and "the patient is no longer free to make rational choices among alternatives" (Pellegrino and Thomasma 1981, 208–209). Also, more recently, authors of empirical studies, who adopt a relational approach to autonomy, have stated that severe illness reduces patients' preferences and capabilities for active participatory roles (Wright 2017; Stajduhar et al. 2010). The second objection is that in such cases action-autonomy is never possible since it is caused by the adaptation of

Fig. 2.1 Types of constraints (action-autonomy undermining factors) in end-of-life decision-making context

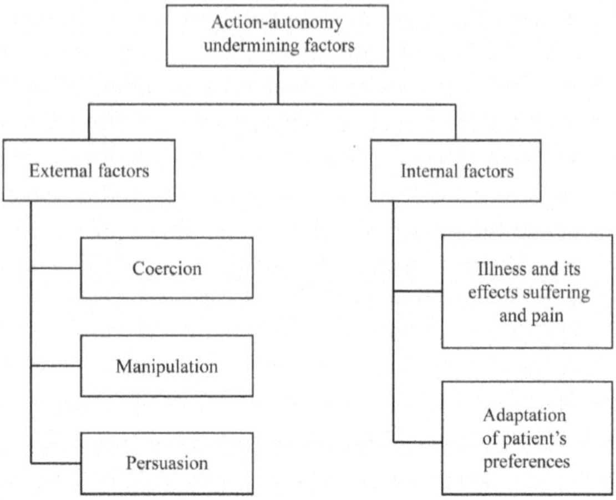

the terminally ill patients' preferences (Colburn 2013, Räikkä and Varelius 2013).[9] Preference adaptation consists in "unconscious altering of our preferences in light of the range of options that we have available" (Colburn 2013, 73). In end-of-life decision-making, preference adaptation starts with a patient facing a devastating set of options—terminal disease and other accompanying conditions that restrict the possibilities available to him. The patient then, by way of an unconscious process, alters his preferences to his new set of circumstances. He adapts "his preferences [in response] to a perceived lack of alternatives" (Colburn 2013, 74). Finally, the process results in the patient "finding himself with the preference to die" (Räikkä and Varelius 2013, 183). Therefore, the terminally ill patient's preference to die is adaptive in character and thus non-autonomous. It is not what the patient truly desires, because it is caused by the lack and unacceptability of alternatives.

The main point of these objections is that in order to be autonomous, the terminally ill patient's choice or preference to die should be caused neither by his illness and its effects (unbearable pain and suffering) nor the lack and unacceptability of alternatives. If his choice or preference to die is caused by either of these factors, it is non-autonomous and hence invalid. Therefore, the final conclusion is that the threat to action-autonomy in end-of-life decision-making does not come only from external influences—coercion, manipulation and persuasion by other individual, but also from internal influences—the very condition of the patient and his adaptive preferences. Taken together, these constraints (action-autonomy undermining factors) in end-of-life decision making can be presented as follows (Fig. 2.1):

[9]Discussion on adaptive preferences stems from John Elster's work on social choice theory (Elster 1983). Besides in discussion on end-of-life decision making, the topic also has figured in the discussion of human enhancement (Räikkä and Varelius 2013) and on female oppression (Nussbaum 2001).

2.6 "Sufficient Autonomy" Standard for Action-Autonomy in End-Of-Life Decision Making

Considering the objections mentioned above, the question that naturally arises in end-of-life decision-making contexts is: to what extent must the terminally ill patient's choice or preference to die be autonomous in order to be valid? Because it is practically impossible that the illness and lack of alternatives will not influence the terminally ill patient's decision, the question is, can his decision to die ever be deemed valid? My answer to this question is in line with the individualistic approach to human autonomy. I think that in cases of end-of-life decision-making we should reject the idea of "perfect autonomy" and adopt a standard of "sufficient autonomy" as a standard for autonomous action. In addition to this, we should set the threshold of "sufficient autonomy" much lower for terminally ill patients than for other patients.

The idea of "perfect autonomy" as a standard for autonomous action comes from Aristotle's concept of deliberate choice or *prohairesis*. For Aristotle, a chosen action is one that is decided upon through a process of deliberation. The process of deliberation requires time, information, a clear head, and highly developed cognitive capacities (Aristotle 2000, 1111b–1112a). His conception of deliberate choice consists of five criteria: (1) the chooser is "competent", (2) he does not choose under coercion, (3) his choice is not the result of manipulation, (4) his choice is not the result of ignorance or mistaken belief, and (5) he does not choose in circumstances that are temporarily distorting. Relatively few acts could perfectly satisfy all of these conditions. Namely, only the actions of normal adult human beings in full control of their deliberative faculties—with calmness and deliberateness, no distracting or unsettling emotions, no neurotic compulsion, no misunderstanding—could satisfy such a standard. To the extent that decision-making is affected by compulsion, misinformation, clouded judgment or impaired reasoning, the choice falls short of perfect autonomy, and is thus regarded as non-autonomous. Therefore, Aristotle's concept of deliberate choice does not present a suitable set of criteria for judging the autonomy of the patient's actions—it sets the standard of action-autonomy too high, and is too inflexible.

The "perfect autonomy" standard has a binary understanding of autonomous action, whereby an agent's actions/choices can be either *perfectly autonomous* or *non-autonomous*. Such a standard is too inflexible for any practical purpose, and its requirements are overly stringent when applied to end-of-life decision-making. It is unsuitable, when it comes to evaluating end-of-life choices, that we regard any decisions that do not qualify as perfectly autonomous as completely non-autonomous. End-of-life decisions require a more nuanced approach. This can be visually represented as follows (Fig. 2.2):

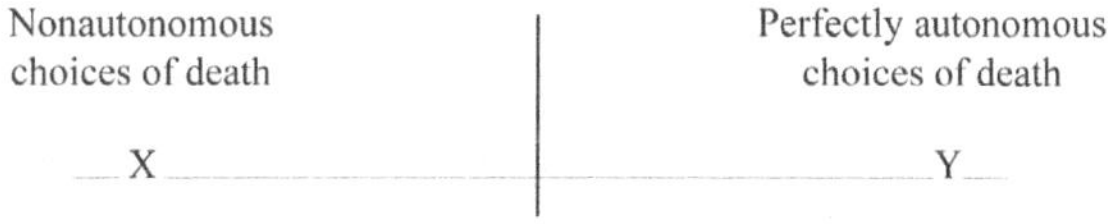

Fig. 2.2 "Perfect autonomy" standard for autonomous action in end-of-life decision-making context

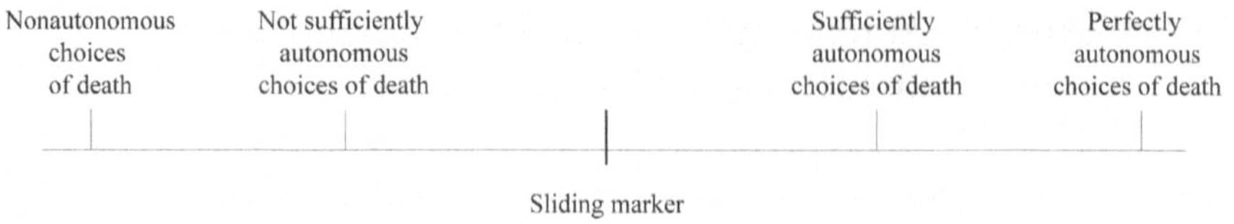

Fig. 2.3 "Sufficient autonomy" standard for autonomous action in end-of-life decision-making context

In order to be applicable to an end-of-life decision-making context, a model for action-autonomy must be adaptable to a variety of medical circumstances and background conditions. Such a standard should be flexible in order to be usable in practical situations, which means that it should recognize the fact that different actions, in different medical circumstances and contexts, should be held to different thresholds of autonomy in order to qualify as worthy of respect. Here, we should avoid a binary standard, and opt instead for a variable conception of autonomous action where the standard is set depending on the nature of the choice and the context in which it is made.

According to this standard for action-autonomy agent's actions/choices can be not just *perfectly autonomous* and *non-autonomous*, but also *sufficiently autonomous* or *insufficiently autonomous.*[10] This model is a suitable means of assessing autonomous action in end-of-life decision-making because it is applicable to a person's acts or choices against specific medical background conditions that may be very restrictive, such as the conditions surrounding end-of-life decision-making. Its application to this context implies that in order to be autonomous, the terminally ill patient's choice to die need not be perfectly autonomous. It is enough that it is sufficiently autonomous, i.e. autonomous enough to be considered as not completely controlled by the illness or the limited and unacceptable alternatives. Therefore, according to this model, we can set different thresholds of autonomy for different contexts and set the threshold of "sufficient autonomy" much lower for terminally ill patients than for other patients. The application of this model to end-of-life decision-making can be presented as follows (Fig. 2.3):

In order to illustrate how the "sufficient autonomy" standard can be applied to the context of end-of-life decisions, imagine the following end-of-life scenario with restricted background conditions: A patient suffers from incurable cancer and knows that he will die within a period of six months. He is experiencing significant pain, anxiety, and depression, and envisages his remaining time as involving constant pain and torment. He comes to the conclusion that his only remaining acceptable alternative is to die by a doctor administered lethal injection, and chooses accordingly. The question is: is his choice autonomous, or since it is unacceptably influenced by internal factors (his illness, unbearable pain, suffering, and preference adaptation due to his suffering), non-autonomous?

[10]For a more extensive statement and defense of this way of construing autonomy see: (Beauchamp and Childress 2013).

If we would use "perfect autonomy" standard for autonomous action, we would see his decision to die as non-autonomous. We would argue that these factors as so distracting that they prevent this patient from seeing a positive perspective of his continued life and a future beyond the scope of his present pain. For example, they prevent him from seeing that the remaining six months might not solely consist of pain and suffering but may also contain some good, the opportunity to see his friends and family. We would conclude that exactly these restricted medical circumstances and patient's distorted view of his situation (seeing just a negative perspective of his continued life) have exerted coercive pressure on him and undermined his autonomy, thus rendering his decision to die essentially non-autonomous.

Advocates of the "perfect autonomy" standard would see his decision to die as non-autonomous. They would argue that his decision is the result of internal factors—his suffering and preference adaptation due to his suffering. They would see these factors as so distracting that they prevent this patient from seeing a positive perspective of his continued life and a future beyond the scope of his present pain. Namely, these influences prevent him from seeing that the remaining six months might not solely consist of pain and suffering but may also contain some good, the opportunity to see his friends and family, for instance. Consequently, advocates of this standard would conclude that exactly these restricted medical circumstances and patient's distorted view of his situation (seeing just a negative perspective of his continued life) have exerted coercive pressure on him and undermined his autonomy, thus rendering his decision to die essentially non-autonomous.

On the other hand, if we tailor our standard for autonomous action to this specific medical context we get quite a different result. The reason for this is that we do not adopt a uniform standard for autonomy, applicable to all situations and circumstances (as advocates of "perfect autonomy" standard would do), but rather tailor our standard to the specific circumstances of this decision. Namely, we take this patient's situation as a given, an appropriate baseline for his specific situation—suffering from incurable illness, experiencing unbearable pain and torment, and seeing the continuation of his life as an unacceptable alternative. Once we tailor our standard to these specific background conditions we can regard this patient's choice to die as autonomous in his setting (which is in its nature very restrictive) because it is sufficiently autonomous, given the circumstances. We can designate his decisions as (sufficiently) autonomous when he is influenced by the internal factors, because we are taking him exactly as he is and not holding him to a standard of autonomy appropriate for a "normal" person, in more unrestricted and less burdensome circumstances.

Therefore, tailoring our standard for autonomous action to various specific medical circumstances enables us to set a different threshold for different medical contexts and to set the threshold of "sufficient autonomy" much lower in cases of end-of-life decision-making, compared to other medical contexts. In the context of, for example, preventive medicine, surrogacy, stem cell research, or artificial insemination, we are dealing with the decisions and choices of patients who do not find themselves in such restrictive circumstances as those who are terminally ill. Accordingly, the thresholds of "sufficient autonomy" for these contexts should be set higher than the thresholds for the end-of-life decisions.

2.7 Conclusion

In this paper, I have examined the individualistic conception of autonomy in the context of end-of-life decision-making. First, I highlighted the importance of making the differentiation between agent-autonomy and action-autonomy. I have then suggested that despite their great influence, internal factors do not undermine action-autonomy in end-of-life decision-making. Although terminally ill patients are hardly ever free from significant internal constraints, the use of a "sufficient autonomy" standard, which is adaptable to their specific circumstances and background conditions, shows that their choice and preference to die can indeed be sufficiently autonomous. We should set the threshold for sufficient autonomy much lower in these cases, in accordance with their far more restricted circumstances. Once we view the assessment of autonomy through this lens, it becomes clear that these decisions, despite the restrictive circumstances under which they are made, should indeed be regarded as autonomous and thus as worthy of respect.

References

Aristotle. 1998. *Politics*, ed. R.F. Stalley and Ernest Barker. Oxford University Press.

Aristotle. 2000. *Nicomachean Ethics*, ed. by Roger Crisp. Cambridge University Press.

Beauchamp, T.L., and J.F. Childress. 1979. *Principles of Biomedical Ethics*, 1st ed. New York: Oxford University Press.

Beauchamp, T.L., and J.F. Childress. 2013. *Principles of Biomedical Ethics*, 7th ed. New York: Oxford University Press.

Benson, P. 2005. Feminist Intuitions and the Normative Substance of Autonomy. In *Personal Autonomy: New Essays on Personal Autonomy and Its Role in Contemporary Moral Philosophy*, ed. J.S. Taylor, 124–142. Cambridge: Cambridge University Press.

Campbell, N. 1999. A problem for the Idea of Voluntary Euthanasia. *Journal of Medical Ethics* 25 (3): 242–244.

Charles, S. 2010. How Should Feminist Autonomy Theorists Respond to the Problem of Internalized Oppression? *Social Theory and Practice* 36 (3): 409–428.

Christman, J. 2004. Relational Autonomy, Liberal Individualism, and the Social Constitution of Selves. *Philosophical Studies* 117: 143–164.

Christman, J. 2009. *The Politics of Persons. Individual Autonomy and Socio-historical Selves*. Cambridge: Cambridge University Press.

Colburn, B. 2013. Autonomy and End of Life Decision: A Paradox. In *Adaptation and Autonomy: Adaptive Preferences in Enhancing and Ending Life*, ed. Juha Räikkä and Jukka Varelius, 69–80. Springer.

Dan-Cohen, M. 2002. *Harmful Thoughts: Essays on Law*. Self and Morality: Princeton University Press.

Donner, W. 2008. Autonomy, Tradition and the Enforcement of Morality. In *Mill's On Liberty: A critical Guide*, ed. C.L. Ten, 138–164. Cambridge University Press.

Dworkin, G. 1988. *The Theory and Practice of Autonomy*. New York: Cambridge University Press.

Dworkin, G. 2015. The Nature of Autonomy. *Nordic Journal of Studies in Educational Policy* 2: 7–14.

Elster, J. 1983. *Sour Grapes: Studies in the Subversion of Rationality*. Cambridge University Press.

Engelhardt, H. T. 2001. The Many Faces of Autonomy. *Health Care Analysis* 9 (3): 283–297.

Faden, R.R., and Beauchamp, T.L. 1986. *A History and Theory of Informed Consent*. New York: Oxford University Press.

Feinberg, J. 1986. *Harm to Self*. Oxford University Press.

Friedman, M. 2003. *Autonomy*. Gender, Politics, New York: Oxford University Press.

Gillon, R. 1985. *Philosophical Medical Ethics*. Chichester: Wiley.

Gillon, R. 2003. Ethics Needs Principles—Four can Encompass the Rest—and Respect for Autonomy should be "First Among Equals." *Journal of Medical Ethics* 29 (5): 307–312.

Gracia, D. 2012. The Many Faces of Autonomy. *Theoretical Medicine and Bioethics* 33 (1): 57–64.

Gray, J. 1996. *Mill on Liberty: a Defence*. Routledge.

Hill, T.E. 1992. The Kantian Conception of Autonomy. In *Dignity and Practical Reason in Kant's Moral Theory*, 79–96. Cornell University Press.

Hill, T.E. 2013. Kantian Autonomy and Contemporary Ideas of Autonomy. In *Kant on Moral Autonomy*, ed. Oliver Svensen, 15–31. Cambridge University Press.

Ho, A. 2008. The Individualist Model of Autonomy and the Challenge of Disability. *Journal of Bioethical Inquiry* 5 (2–3): 193–207.

Hustey, F.M., and S.W. Meldon. 2002. The Prevalence and Documentation of Impaired Mental Status in Elderly Emergency Department Patients. *Annals of Emergency Medicine* 39 (3): 248–253.

Ikonomidis, S., and P. Singer. 1999. Autonomy, Liberalism and Advance Care Planning. *Journal of Medical Ethics* 25 (6): 522–527.

Kant, I. 2002. *Grounding for the Metaphysics of Morals*, ed. Allen W. Wood. New Haven: Yale University Press.

Kemp, P. and J.R. Dahl. 2008. The Barcelona Declaration: Towards an Integrated Approach to Basic Ethical Principles. *Synthesis Philosophica* 23 (2): 239–244.

Krishna, L.K., D.S. Watkinson, and N.L. Beng. 2015. Limits to Relational Autonomy—the Singaporean Experience. *Nursing Ethics* 22 (3): 331–340.

Mackenzie, C., and N. Stoljar. 2000. *Relational Autonomy: Feminist Perspectives on Autonomy, agency, and the Social Self*. New York: Oxford University Press.

May, T. 1994. The Concept of Autonomy. *Philosophical Quarterly* 31 (2): 133–144.

McLeod, C. 2002. *Self-Trust and Reproductive Autonomy*. Cambridge, MA: MIT Press.

Meyers, D.T. 2002. *Gender in the Mirror: Cultural Imagery and Women's Agency*. New York: Oxford University Press.

Mill, J. S. 2009. *On Liberty*. The Floating Press.

Nussbaum, M. 2001. Adaptive Preferences and Women's Options. *Economics and Philosophy* 17 (1): 67–88.

O'Neill, O. 2002. *Autonomy and Trust in Bioethics*. Cambridge University Press.

Oshana, M. 2016. *Personal Autonomy in Society*. London: Routledge.

Pellegrino, D. Edmund., and C. David. Thomasma. 1981. *A Philosophical Basis of Practice: Toward a Philosophy and Ethic of the Healing Professions*. New York: Oxford University Press.

Pisani, M.A., L. McNicoll, and S.K. Inouye. 2003. Cognitive Impairment in the Intensive Care Unit. *Clinics in Chest Medicine* 24 (4): 727–737.

Räikkä, Juha and Jukka, Varelius. (eds.). 2013. *Adaptation and Autonomy: Adaptive Preferences in Enhancing and Ending Life*. Springer.

Raymont, V., W. Bingley, A. Buchanan, A.S. David, P. Hayward, S. Wessely, and M. Hotopf. 2004. Prevalence of Mental Incapacity in Medical Inpatients and Associated Risk Factors: Cross-Sectional Study. *Lancet* 364 (9443): 1421–1427.

Richardson, H. 2001. Autonomy's Many Normative Presuppositions. *American Philosophical Quarterly* 38: 287–303.

Schaeffer, M.H., D.S. Krantz, A. Wichman, H. Masur, E. Reed, and J.K. Vinicky. 1996. The Impact of Disease Severity on the Informed Consent Process in Clinical Research. *American Journal of Medicine* 100 (3): 261–268.

Schneewind, J.B. 2013. Autonomy after Kant. In *Kant on Moral Autonomy*, ed. Oliver Svensen, 146–69. Cambridge University Press.

Secker, B. 1999. The Appearance of Kant's Deontology in Contemporary Kantianism: Concepts of Patient Autonomy in Bioethics. *Journal of Medicine and Philosophy* 24 (1): 43–66.
Stajduhar, K., L. Funk, E. Jakobsson, and J. Öhlén. 2010. A Critical Analysis of Health Promotion and 'Empowerment' in the Context of Palliative Family Caregiving. *Nurs Inquiry* 17 (3): 221–230.
Taylor, J.S. 2005. Introduction. In *Personal Autonomy: New Essays on Personal Autonomy and Its Role in Contemporary Moral Philosophy*, ed. Taylor James Stacey, 1–29. Cambridge: Cambridge University Press.
Taylor, J.S. 2009. *Practical Autonomy and Bioethics*. Routledge.
The National Commission for the Protection of Human Subjects of Biomedical and Behavioral Research. 1978. *The Belmont Report: Ethical Principles and Guidelines for the Protection of Human Subjects of Research*. Washington, D. C.: United States.
Turner, N.P. 2013. The Absolutism Problem in On Liberty. *Canadian Journal of Philosophy* 43 (3): 322–340.
World Medical Association. 2013. World Medical Association Declaration of Helsinki: Ethical Principles for Medical Research Involving Human Subjects. *JAMA* 310 (20): 2191–2194.
Wright, M.S. 2017. End of Life and Autonomy: The Case for Relational Nudges in End-of-Life Decision-Making Law and policy. *Maryland Law Review* 77 (4): 1062–1141.

Milijana Djerić is a Research Associate at the Faculty of Philosophy at the University of Belgrade, Serbia. She obtained her Ph.D. degree in 2018. The title of her dissertation was "*Ethical Dimensions of the Distinction Between Passive and Active Euthanasia.*" Her research interests include ethics and applied ethics, with special focus on bioethics. Her recent publications include: "Bratman's and Cavanaugh's Analysis of the Intention: Solution to the Problem of the Closeness of the Doctrine of Double Effect" *Theoria* 61/4, 2018, 137–146; "Euthanasia in Serbia: In Need of Enhanced Debate and Regulation", *JAHR: European Journal of Bioethics* 7/13, 2016, 55–60; "Do We Have a Duty to Help Those Who Are Dying of Starvation: Some Implications of Rachel's Thesis About Moral Equivalence of Killing and Letting Die," *Felix Romuliana: The International Scientific Philosophical School, Collection of Papers 2007-2015*, 2015, 590–601; "The Doctrine of Double Effect and Terminal Sedation," *JAHR: European Journal of Bioethics* 5/10, 2014, 255–261; and "Euthanasia: The Conceptualization of the Problem and Important Distinctions," *Filozofija i društvo* 24/2, 2013, 255–263.

Chapter 3
Personal Identity and Self-Regarding Choice in Medical Ethics

Lucie White

Abstract When talking about personal identity in the context of medical ethics, ethicists tend to borrow haphazardly from different philosophical notions of personal identity, or to abjure these abstract metaphysical concerns as having nothing to do with practical questions in medical ethics. In fact, however, value of and moral authority for respecting a patient's self-regarding decisions can only be made sense of if we make certain assumptions that are central to a particular, psychological picture of personal identity, namely, that patients will remain psychologically connected to a certain degree with their future selves.

3.1 Introduction

Philosophical concerns with personal identity have long formed a central part of debate on certain bioethical issues such as abortion (McInerney 1990; Oderberg 1997; Warren 1977) and advance directives (Buchanan 1988; DeGrazia 1999). But these considerations have received renewed attention in the context of the personality changes potentially induced by deep brain stimulation (Jecker and Ko 2017; Müller, Bittlinger and Walter 2017; Nyholm and O'Neill 2016). The focus of this paper is to point out some problems that arise in some of these recent discussions of personal identity and medical ethics—particularly, when it comes to understanding the importance of respecting self-regarding choice in medical contexts. It is widely agreed that it is very important to respect a patient's self-regarding decisions in medical ethics. I argue that once we delve into this, we will see that some of the value we place on making our own decisions relies on an assumption that is central to a particular picture of personal identity—that we will remain psychologically connected to our future selves to an adequate degree. This means, roughly, that psychological elements such as beliefs, desires, and values remain stable over time, or connected in certain ways to (e.g. endorsable on the basis of) previous psychological

L. White (✉)
Institute of Philosophy, Leibniz Universität Hannover, Hannover, Germany
e-mail: lucie.white@philos.uni-hannover.de

M. Kühler and V. L. Mitrović (eds.), *Theories of the Self and Autonomy in Medical Ethics*, The International Library of Bioethics 83,
https://doi.org/10.1007/978-3-030-56703-3_3

elements. Recognising that psychological connectedness is important for this purpose does not require us to make any metaphysical claims about what a person is through time—we do not, more specifically, need to endorse a psychological-connectedness-based metaphysical account of personal identity. But work in medical ethics based on non-psychological theories of personal identity runs into problems insofar as it is unable to recognise the importance of psychological connectedness.

I will draw this out as follows: first, I will delve into the moral basis for according so much importance to respecting self-regarding decisions in medical ethics, with the help of work from T.M. Scanlon. I will look at the various reasons that we generally have to value making our own choices, showing that part of the reason we value making choices with a significant, long-term impact relies on an underlying assumption that the agent making the decision will be psychologically connected to her future self to an adequate degree. Although a focus on psychological connectedness is the basis of a psychological or neo-Lockean theory of personal identity, I will argue that we don't have to make any metaphysical commitment about what the "self" is through time to recognise the importance of psychological connectedness as the basis for self-regarding decisions. I will situate this within a more general account connecting the reasons we have to value making our own choices with concerns typically discussed in the context of personal identity (Sect. 3.2).

With this in hand, I will turn to two treatments of personal identity in connection with medical ethics, which, I will argue, run into problems due to a failure to recognise the importance of psychological connectedness. First, I will look at a discussion of advance directives, and how personality changes might impact their authority. I will argue that the moral authority of advance directives is unavoidably based, in part, on an assumption of psychological connectedness. I will show that this comes apart from metaphysical questions about personal identity, and that it can explain our reaction to certain difficult cases. I will argue that this recognition doesn't require us to abandon advance directives where psychological connectedness is disrupted (they may be justified on other grounds) but that we have to recognise the relevance of psychological connectedness to questions about the value of advance directives and the circumstances under which they should hold (Sect. 3.3).

I will then turn to a rather different problem—the problem of making medical decisions that may involve drastic psychological change, brought on by deep brain stimulation. I will argue that without an underlying basis of psychological connectedness motivating our decisions and delineating what counts as wanted (or unwanted) personality change, we have no basis upon which decisions can be made. This means that we cannot capture or understand what is of concern to the patient when she expresses fear or reservations about personality change brought on by deep brain stimulation. I will show that psychological connectedness does not require that the psychological attributes of the person remain static through time—certain changes are compatible with psychological connectedness (Sect. 3.4). These two cases show that we can't completely understand the *value* of making self-regarding decisions over time, and thus the source of *authority* of these decisions, and what is of *concern* to the patient when making these decisions, without engaging with considerations of psychological connectedness. Psychological connectedness is inextricably intertwined with the value of self-regarding choice.

3.2 Psychological Connectedness and Long-Term Decision-Making

Although it is widely agreed that respect for self-regarding choices made under certain conditions (often referred to as "competent" or "autonomous" choice) is a vital or even the preeminent ethical value in medicine (Faden and Beauchamp 1986; Gillon 2003; Taylor 2009), there is sometimes limited reflection about just why it is that we value making our own choices so much. In order to understand the link between self-regarding choice and psychological connectedness, we will need to delve into this. A useful tool here will be Scanlon's account of the reasons we have to value making our own choices (1998). Although Scanlon presents this theory in a rather different context, we see similar concerns echoed in the literature on medical ethics. Scanlon suggests that there are three reasons that most of us have to value making our own choices, under the right conditions.

Instrumental Value: The most obvious reason that we generally have to value making our own choices is because this has instrumental value to us—making our own choices often appears to be the best way of achieving our own ends or goals, whatever they are. Scanlon gives the example of choosing what to eat at a restaurant. Given my own knowledge of my tastes and values, I might believe that choosing for myself is the best means of achieving my end of enjoying a satisfying meal (or whichever other end I wish to pursue in making a choice). Whether making my own choice does indeed present the best means to achieving this end, however, is clearly dependent on circumstance. If I cannot read the language that the menu is written in, or my capacities for discernment are clouded by hunger, I might not be in a good position to make this choice. The instrumental value of my choice might be improved by additional information, or attempts to mitigate influence affecting my judgment. Alternatively, it might turn out that someone else is in a better position to make the choice that best allows me to achieve my own ends. The instrumental value of choice is often emphasised as a significant reason to value making our own choices in medical contexts (Buchanan and Brock 1989; Glover 1977), based, crucially for our purposes, on "the assumption that a competent person is the best judge of his or her own interests." (Buchanan and Brock 1989, 153). Some theorists argue that the value of autonomy in medical ethics is purely instrumental (Taylor 2009; Varelius 2006).

Representative Value: Scanlon, however, enumerates two further types of non-instrumental reasons we typically have to value making our own choices. Sometimes, he suggests, we value making our own choices, even when they are not the best means of achieving our desired outcome, because of what our choices then express or represent. We might want to choose our own gift for our spouse, or the furnishings in our home, even if someone else could do a better job, because of what the choice represents—our tastes, for example, or what we value. We might also make and value choices for representative reasons in medical contexts (Buchanan and Brock 1989, 39; Glover 1977, 81).

Symbolic Value: The last reason to value making our own choices, as listed by Scanlon, is because the ability to make these choices carries a certain social weight. In situations in which people are normally expected to make their own choices, being denied this ability can be seen as reflecting a judgment that a person lacks "the standing normally accorded to an adult member of society" (1998, 253). Being denied the ability to make these choices may be experienced as demeaning, showing insufficient respect for a person's capacities, competence or worth. It is this dimension which is most closely connected to professions that persons have a "right" or "authority" to make certain of their own choices (even when we know that others could choose better), particularly choices concerning how to dispose over one's body (Dworkin 1994; Feinberg 1986). Recognizing a person's authority to make these decisions is sometimes regarded as essential to recognizing the dignity and worth of persons (Darwall 2006). We see similar sentiments in medical ethics, reflected in a view of making one's own choice as a valuable means of exercising and exhibiting the "capacities for deliberation, choice and action that normal humans possess" (Buchanan and Brock 1989, 38).

This is not necessarily an exhaustive account of all the reasons we might have to value making our own choices,[1] and these reasons, in practice, will often coexist and can be difficult to distinguish.[2] However, this gives us a good basis upon which we can start to draw in concerns usually thought to belong to the philosophical field of personal identity. Theories of personal identity are often concerned with identifying just what is it that makes a person the same person over time. This is often referred to as the question of persistence or reidentification. It is this time dimension that invokes similar issues when it comes to self-regarding choice in medical ethics. We can make all sorts of self-regarding decisions in a medical context. Some of these decisions are relatively low-stakes or easily reversible (like deciding to take certain types of medications). But other decisions (like, for example, the decision to undergo an amputation) are high-stakes, and have a long-term impact. This means that when you make such a decision, this will have an impact on your distant-future self. You are making this decision on behalf of your distant-future self as well as your present self. This is even more clearly the case when you issue an advance directive—here, you are making a choice purely on behalf of your future self.

[1]Göran Duus-Otterström (2011) lists several additional potential categories of reasons, and it might be the case that certain people have their own idiosyncratic reasons for valuing choice.

[2]Considerations linked to the representative and symbolic value of choice are often conflated to a certain extent in the literature on medical ethics (see Buchanan and Brock 1989, 38–9; Glover 1977, 81). This is likely because the discussion is often framed in terms of discerning whether choices have value beyond the obvious instrumental dimension.

The instrumental value of choice, as we have seen, is based upon the idea that a patient has particular insight into his own tastes, preferences, values and interests, so that he is in a position to make self-regarding decisions with particularly high instrumental value, as long as he is making the decision under sufficiently good circumstances (in possession of sufficient information, sufficiently free of controlling influences, etc.). When we are making decisions with long term impact, we need to add an additional condition here—in order for your decision to be sufficiently likely to have high instrumental value for your distant-future self, you need to be sufficiently psychologically connected to your distant-future self, so that your insight into your interests also provides special insight into his interests.[3] That is, the values and preferences that you wish to serve or satisfy in making the decision must be shared by or sufficiently similar to the values and preferences of your distant-future self, for your future self also to regard this decision as of high instrumental value. If there is a significant shift in one's psychological makeup in the interim, there may be reason to doubt that you are in the best position to make the choice that is likely to have the highest instrumental value. The balance might, in some cases, swing in the other direction, and suggest that instrumental value might best be pursued through having someone else make this decision, just as in cases where the circumstances under which you are making the choice cannot be made sufficiently good.

Similar remarks apply to choices with representative value—if you make a choice in order to express your commitment to something that you value, and later come to repudiate this value due to psychological changes, your choice will have negative representative value from the perspective of your future self—he will be likely to regret having made this decision in the past. The balance here is slightly different, because you need to make your own choice for it to have representative value—a choice made by someone else cannot express something about yourself. Someone else, that is, cannot make a choice on your behalf that has higher representative value. However, it is clear that when it comes to long-term decisions, and in the case of drastic personality change, the representative value of making your own choices may be diminished.

Such concerns do not affect the symbolic value of choice—tied to the right to make your own decisions regardless of outcome. I will make further comments about how the symbolic value of choice intertwines with considerations of personal identity in the subsequent section. From the above considerations, however, we can see that psychological connectedness is an important underpinning of some of the sources of the value of making high-impact, long term decisions (though psychological disruption may not affect all of the reasons we have to value making our own choices). The importance of the instrumental and representative value of choice should not

[3]This, of course, doesn't *guarantee* that the decision you make will have high instrumental value for your distant-future self, nor does it rule out the possibility that someone can make a decision with high instrumental value for a future person despite a lack of psychological connectedness. Just as with possession of knowledge, or freedom from coercive elements, psychological connectedness simply serves to increase the chances that the decision will have high instrumental value, to the degree that we can count on the decision having instrumental value if made by a particular person in particular circumstances.

be underestimated—it is only through reference to these values that we can explain why informing a patient, or ensuring that she is free of controlling influence, is so important in medical ethics—these measures put the patient in circumstances that better allow her to align her actions with her goals, or the values she wants to express through her choice.

This notion that certain psychological factors should remain stable over time is not alien to medical ethics; in their influential accounts of competence to consent to treatment in medical ethics, both Dan Brock (1991) and Thomas Grisso and Paul Appelbaum (1998) maintain that high-stakes treatment decisions should be scrutinised in order to ascertain that the beliefs and values underlying them are *enduring*. Though far from a perfect measure, the stability of beliefs and values underlying a treatment decision might provide the best indicator we have that these values will persist into the future, and thus that the treatment decision will continue to be well-regarded by the patient into the future. This, of course, is compromised in cases where we have reason to suspect that a disruption in psychological connectedness might ensue.

3.2.1 A Metaphysical or Normative Commitment?

In the previous section, you may have noticed that I avoided adopting an explicitly psychological picture of personal identity—I have assumed that although a disruption in psychological connectedness might diminish the value of making our own decisions, it still makes sense to talk about a single person existing prior to and after a personality shift that leads to a loss of these connections. In this section, I will connect what I have said in the previous section to the debate concerning different theories of personal identity. To draw this out, let's first take a step back to outline just what a psychological theory of personal identity, is, and what its main competitors are. This will also lay the groundwork for my critique of bioethical work based on non-psychological theories in the subsequent sections.

The problem of personal identity I am focusing on here, as noted above, is the problem of reidentification or persistence—what allows us to say that someone is the same person at two different points in time, and, on the opposite side of the same coin, what type of changes can cause a person to cease to exist? Psychological or neo-Lockean theorists suggests that the key is direct *psychological connections* that endure over time. We are linked to our previous selves through memories of past experiences, actions motivated by past intentions, enduring beliefs, desires, values, and other psychological elements. If enough psychological connections exist between two people at two different points in time, we can say that they are the same person. Some theorists prefer to focus more on *psychological continuity* than connectedness (see Schechtman 2010) holding that it is not direct connections that are important in establishing identity through time, but rather an overlapping and unbroken chain of psychological connections, stretching through time like a rope, allowing for the retention of personal identity where no direct connections remain, while some theorists think that both connectedness and continuity are important (Parfit 1984). Either way, some sort of drastic shift in our psychological contents, a fundamental change in our values, beliefs, desires, created by, for example, a religious epiphany, deep brain stimulation, or dementia, might be enough to disrupt connectedness (and thus also continuity), leading us to say that the resulting person is not the same as the person who came before. The previous person has ceased to exist, and a new person has (perhaps) come into existence. Just how many psychological connections we need to establish continued identity is a matter of debate—some prefer to set the threshold quite high, requiring many enduring or overlapping connections, while others think that a lower threshold is preferable (Buchanan and Brock 1989). Such a theory is a metaphysical in nature, it aims to tell us what the fundamental nature of persons is, what is it for a person to exist, or not to exist (see Nyholm 2018). But psychological theories of personal identity are also closely tied to practical concerns—this theory of personal identity, according to adherents, can tell us who to hold responsible for past crimes, or who is the appropriate object of prudential concern (Schechtman 2010).

The main competing metaphysical theory of personal identity argues that it is rather *biological continuity* that is the key to personal identity—as long as you continue to be the same biological organism, with the same body (or maybe a specific part of the body, such as the brain stem), you are the same person. Adherents to this type of theory are generally motivated by metaphysical rather than practical concerns—problems to do with picking out a single person over time (Williams 1957), or establishing how we are identical to the foetus from which we developed (Olson 1997). In fact, the most prominent contemporary advocates of biological continuity argue that it is purely metaphysical in nature—that uncovering what we are at the most fundamental level, and the conditions for our persistence, does not have much to do with practical ethical questions—to address these, we still need to appeal to psychological continuity (Olson 1997, 69-71, DeGrazia 1999, 338–9).

A third popular theory of personal identity, which has been utilized extensively in recent work on medical ethics, particularly in the context of deep brain stimulation (see Baylis 2009; Jecker and Ko 2017; Nyholm and O'Neill 2016) is the *narrativetheory*. This theory is not metaphysical in the sense of the other two—rather than focusing on what a person is at the most fundamental level, and how to reidentify a person at a later point in time, this theory asks how we can attribute actions, beliefs, values, character traits and so on to a given person. The answer to this, according to narrative theorists, is that these traits are attributable to the person insofar as they are incorporated into an intelligible and coherent narrative that gives shape and meaning to the individual's life (Schechtman 1996). This theory is intended to account for practical concerns such us as when is it appropriate to attribute responsibility to someone.

We can now turn to the connection between theories of personal identity and the value of choice. Although the instrumental value of choice is contingent on a person remaining sufficiently psychologically connected to her future self, so that she is in the best position to make decisions on behalf of her future self, this does not require any metaphysical commitment about what the self is on its most fundamental level. We could endorse any of the theories of personal identity outlined above, while still maintaining that psychological connectedness is important for this specific purpose—it is a condition of ensuring that decisions are sufficiently likely to have sufficient instrumental and representative value. Let's imagine, for example, that we are biological continuity theorists—we simply accept that persons *always retain* their identity throughout the entire course of their bodily lives, and that no psychological change could possibly disrupt continued identity. This does not give us a reason to expect that a decision will have instrumental value for your future self in the absence of psychological connectedness, or that you are necessarily in a particularly good position to make a decision for your future self.

At the other end of the spectrum, we might be sceptics about personal identity, believing that there simply is no metaphysical entity that exists through time, and remains the same (see Parfit 1984), and thus that we *never retain* identity through time. This does not undermine the fact that a person who is very psychologically similar to a person in the future will be well placed to make decisions with high instrumental value for this future person. The instrumental value of choice comes apart from our metaphysical commitments about personal identity. It is rather a normative commitment—a significant reason that we value making our own decisions is because this has instrumental value to us, but we can only expect that choices will have this instrumental value if we are psychologically connected to the person who made the decision. The importance of psychological connectedness comes from the fact that it underpins something we value—the instrumental value of choice.

The representative value of choice is more complicated—for a choice to have representative value, *you* need to have made it, not someone else who is psychologically similar to you. But a metaphysical theory of personal identity does not seem to help much in whether you can or will regard a past decision as expressing or representing something about yourself—it does not seem to be the case, for example, that you will necessarily regard the choice as your own if the biological continuity theory of personal identity is the "true" metaphysical theory that best describes the fundamental nature of what exists in the world, and you share a body with the person who made the choice. Perhaps the narrative theory of identity is of some use in approaching this—as long as you regard a past choice as your own, you attribute it to yourself, and it can continue to have representative value for you. Despite these complications, however, we can still recognise that choices are likely to have diminished representative value if you undergo a shift in what you value and believe—the attributes that you wanted to express in your decision might be ones that you later come to feel indifferent about or even repudiate. So psychological connectedness is certainly an important underpinning of this value of choice, and this aspect can be recognised while remaining neutral about what the persistence conditions of persons are (although it might turn out that the representative value of choice can be diminished or nullified in other ways, if you, for whatever reason, can no longer regard a past choice as your own).

To settle the question of symbolic value, on the other hand, it does seem like we will need a metaphysical answer to the question of persistence. If we see symbolic value as linked to the principle that you have a right to decide for yourself, we need to know whether a future person is *you* to know whether you have the right to decide for this person on these grounds.[4] If we imagine, for example, that the "true" theory of the self is psychological continuity, then it would be a misapplication of this principle to make decisions on behalf of a future person who has undergone drastic psychological change, and is thus no longer you. Scepticism about personal identity will also have serious ramifications here—if the boundaries of the self do not exist and can't be clearly defined, it might be the case that this principle does not have applicability through time.

3.3 The Source of Value: Biological Continuity and Advance Directives

With this groundwork in place, we are now in a position to turn to problems with work in medical ethics that rests on problematic assumptions about personal identity. As we've seen in the previous section, we can recognise that an underlying assumption of psychological connectedness is important in ensuring that choices have instrumental and representative value, no matter what position we take on personal identity. However, problems can arise when bioethicists endorse a theory of personal identity which is unable to recognise or accommodate the importance of psychological connectedness.

Take, for example, Sabine Müller, Merlin Bittlinger and Henrik Walter's work on personal identity and potentially personality-altering neurosurgeries (2017). Müller and colleagues are concerned about the role that metaphysical ideas about personal identity play in bioethical discussions. They are particularly concerned that invoking certain theories of personal identity (mainly psychological, but also narrative theories) might be used to undermine advance directives, which they take to be of central importance in ensuring that patients have the "right" to "exercise their autonomy" (2017, 300). They instead endorse a biological continuity-based theory of personal identity, not for metaphysical reasons, but because this theory guarantees identity between a person making the decision and a person at a later stage who has lost many of their previous capacities, someone, for example, with end-stage dementia, or who is in a coma, ensuring that an advance directive will retain its moral and legal force. They argue that we should keep discussion of psychological continuity or connectedness out of bioethical discussions altogether. Ultimately, they contend, "the application of psychological continuity theories of personal identity on these neuroethical issues—particularly adapted versions of Parfit's theory and its metaphysical implications—is of limited use." (2017, 308).

Müller and colleagues' desire to retain the authority of advance directives is certainly understandable. But the route which they take to defend advance directives is beset with several problems. First, as we have seen, these concerns about

---

[4] *On these grounds* is important here; Allen Buchanan and Dan Brock mount a compelling argument that advance directives can be justified after significant personality change (that they take to sever identity) on other grounds (1989, 163–5), which we will explore below.

psychological connectedness are not purely metaphysical. Although they echo the metaphysical debate about personal identity, the grounds upon which I have invoked psychological connectedness are purely normative. We value making our own decisions, in part, because we believe that we can decide in a way that will have the most instrumental and representative value for ourselves. In cases where drastic personality change greatly affects the things that we value, we can no longer make this assumption. This may change the balance of values in these cases. And acknowledging this has nothing to do with metaphysical concerns.

In order to illustrate this, let's turn to a difficult case considered by bioethicists Allen Buchanan and Dan Brock, in their discussion of personal identity and advance directives. Buchanan and Brock accept, for the sake of argument, a psychological continuity-based concept of personal identity, and consider the consequences for advance directives. They argue that the degree of psychological continuity required for continued identity should be quite low. A patient with severe dementia, severe brain damage, or in a coma will fall below the threshold required for continued identity, but someone with personality change resulting from moderate dementia or brain damage, for example, should still be considered to be the same person. They imagine a case in which, according to this picture, personal identity is retained:

> An apparently valid advance directive specifies that no life-sustaining treatments, including antibiotics, are to be used on the individual if he or she suffers a loss of cognitive function. The patient does suffer a serious loss of cognitive function due to a discrete neurological injury, such as a stroke. He or she is now a moderately mentally handicapped person, but clearly a person – and indeed the same person – nonetheless. Moreover, the patient is quite happy, and wants to live. Then the patient develops a life-threatening pneumonia. Following the advance directive will result in the easily avoidable death of a happy and reasonably healthy, although mentally handicapped person. Suppose, also, that he or she is judged not to be competent to rescind the advance directive. (1989, 188–9)

This case, they argue, presents a challenge for the absolute authority of advance directives, and makes the exercise of paternalism in this case tempting. I believe the picture of the value of self-regarding choice I have given above can explain our reluctance to enforce an advance directive in these circumstances. The changes in the patient's psychological make-up means that at the time she issued the advance directive, she was not in a good position to imagine which choices would have the most instrumental value for her in her current state. Perhaps she could not imagine that she could possibly be happy, or have a life worth continuing, in this state. Now we are faced with a patient who is clearly happy. The advance directive recommends a course of action which seems so clearly to be against her own happiness and interests, subjectively conceived. It is this, I believe, which accounts for our reluctance in this case, and a possible weakening of the moral authority of an advance directive under these circumstances. And this holds even if we accept that the patient is definitely the same person as the person who issued the advance directive.

Does this mean that we should refuse to recognise the authority of advance directives under these circumstances? Rebecca Dresser, in her consideration of a similar situation, argues that we should not honour advance directives in such cases (2006).

On the other hand, attempts to institute limits on advance directives along these lines might lead to them being ignored on a wider scale than appropriate in favour of medical paternalism (Buchanan 1988), so we might have good pragmatic reasons for enforcing them in these cases. *If* we take for granted that identity is retained here, we also still have the symbolic dimension of choice to consider. Karsten Witt suggests we should deal with personality change by having the patient consider her interests from the perspective of someone who is in the post-change position, as well as her current position (2017). Although it doesn't seem that the patient is necessarily in a particularly good position to recognise her interests post-personality change (compared to others that may have more insight into the resulting condition), this might provide a possible means of retaining the symbolic value of choice while enhancing the instrumental value of choice to some degree, and guarding against unwarranted paternalistic creep.

It is beyond the scope of the paper to settle this issue—the point here is that we cannot understand the source of the value of advance directives, and thus the reason that we regard them as having moral authority, without considering psychological connectedness and the instrumental value of choice. Simply asserting that we should endorse a particular picture of personal identity does not allow us to avoid the issues that arise when psychological connectedness is disrupted.

Much of what I have said above may not apply to the most extreme cases considered by Müller and colleagues; cases, for example, in which the patient is in an irreversible coma. The problem here is not that the patient may have changed in ways that may lead to her no longer regarding her previous decisions as prudent or conducive to her happiness and interests, but that she has lost the ability to consider these choices at all. Some (psychological continuity) theorists see this problem not so much as involving the destruction of one person and the creation of another, but rather a loss of personhood altogether. Many advance directives will involve circumstances such as these. How then, should we approach these cases? Is there any way to recognise their authority without endorsing a biological-continuity-based picture of personal identity?

Buchanan and Brock give us an alternative means of approaching these problems. As we have seen, they assume a psychological theory of personal identity, with a low threshold of continuity required to retain identity. It is an unavoidable consequence of their theory that personhood is severed in these extreme cases. However, they do not take this to undermine the validity of advance directives. Just as we generally take a person to have interests surviving death (they might have an interest in taking care of their family, for example, that we honour after death through executing their will), they argue, we can take a person to have interests surviving permanent unconsciousness, or severe brain damage that severs identity (1989, 163–4). These include what will happen to the body that they used to inhabit, what kind of resources should be expended to keep it alive and at what emotional and financial cost to their family, and so on. Such interests may need to be weighed against the current "experiential interests" of the person in a severely brain damaged state, for example, but they can and should play a role in what happens to this person.

Müller and colleagues' suggestion to keep abstract metaphysical concerns about psychological continuity out of bioethical discussion therefore misses an important dimension of why we value making our own choices and what may cause this value to be diminished. These concerns about psychological continuity are inextricable from the value that we assign to making our own choices, if we believe that instrumental value is an important source of the value of choice. In other words, we cannot understand the value of advance directives independently of psychological connectedness over time. The assumption of psychological connectedness over time plays a key role in why we think "the exercise of future-directed personal autonomy" (Müller, Bittlinger and Walter 2017, 303) is important, and should override other concerns such as a concern for wellbeing. Furthermore, the imperative to respect advance directives in extreme cases can be justified on other grounds, without maintaining that identity must be regarded as inhering in the continuity of the body.

But a defender of Müller and colleagues might argue that I have missed the point here—that when they talk about autonomy, they do not see the importance of this as grounded in instrumental concerns or interests. After all, they talk about a "right" to exercise one's autonomy (2017, 300). Might it be the case, then, that they see the importance of autonomy as rooted in its symbolic value—that we should simply be allowed to make self-regarding decisions, and that this has nothing to do with their instrumental value? As we have seen above, this route creates another set of problems for Müller and colleagues. Namely, we need to know that a person *is the same person* as the person who issued an advance directive, before we know that it is appropriate to enact this advance directive on the grounds that the person has a right to decide for himself. We cannot avoid metaphysical concerns in working out how this right can be exercised. Prominent biological continuity theorist Eric Olson goes even further than this, arguing that even though biological continuity explains what we are at the most fundamental level, this cannot be used to make *any* ethical inferences (1997, 69–71; see also Schechtman 2010, 276–7; DeGrazia 1999, 338–9). That is, even if we accept that we are our bodies, additional argument might be required to claim that we have a right to decide for ourselves in this sense. Again, Buchanan and Brock's suggestion provides us with a more promising way of maintaining that you should be able to make decisions on behalf of your future body, whether you share identity with this body or not.

3.4 Capturing What We Value: Narrative Identity and Deep Brain Stimulation

A second, more popular strand of argument that runs into problems is based on a narrative picture of identity, which is often used to make sense of the personality changes resulting from deep brain stimulation, and how they are regarded and approached by the patient (see e.g. Baylis 2009; Blustein 1999; Jecker and Ko 2017; Nyholm and O'Neill 2016). Where the previous section dealt with issues that arise after a disruption in psychological connectedness has already taken place, this section will deal with making decisions about potential psychological changes that are yet to come. How do we understand what concerns the patient in making these decisions, and put forward a picture that captures what he values, and the basis upon which he makes these decisions? I will argue here that we can only make sense of what the patient values when making decisions involving personality change if we appeal to psychological connectedness—in making this decision, the patient aims to retain certain psychological connections, and it is the disruption of these connections that concerns the patient when he expresses concern about personality alterations.

Jecker and Ko (2017) provide us with a good example of a defence of a narrative approach to identity as a means of understanding the personality changes brought on by deep brain stimulation. Jecker and Ko want to work out which changes might be experienced by patients as problematic, and get to the heart of what concerns patients are expressing when they are worried that personality changes will mean that they don't 'survive' the intervention - that they will become, in some sense, a different person. As we have seen above, narrative accounts of identity are not metaphysical—they are concerned with what traits are properly attributable to a person. Jecker and Ko see this as a strength of narrative identity in this context—they want to capture and focus on what matters to the patient, rather than the metaphysical conditions for persistence. They suggest that insofar as psychological changes can be incorporated into a patient's "self-told story of his or her life" (2017, 160), these new traits can be attributed to this person, and this captures what patients mean by survival in this context. As long as these changes can be weaved into the patient's own narrative, the patient need not fear that she will become a different person. This understanding of the self, Jecker and Ko argue, provides a good basis for making these medical decisions.

Narrative identity, as a guide for decision making, comes apart from theories that emphasize the importance of one's stable values, beliefs, and other psychological traits. Though these factors might form important reference points in one's narrative, narrative theorists emphasize that making a decision that fits your narrative is not the same as deciding on the basis of your settled values and beliefs (Blustein 1999). How then, do we decide which changes might be incorporated into our own life story? There are certain constraints that narrative theorists place on the construction of a narrative—some emphasize "correctness" (Jecker and Ko 2017) or a "reality constraint" (Schechtman 1996); the requirement that narratives must not contradict "basic, well-established, matters of observable fact" (Schechtman 1996, 121). Some theorists also contend that a narrative needs to be coherent to a certain degree, although it's not quite clear how stringent this standard should be (Blustein 1999).

So, there are some constraints on which of our possible decisions might fit our narrative. But this doesn't give us much guidance concerning which decisions we should make. Narrative identity is compatible with drastic change (Baylis 2009), and several alternatives are likely to fit a possible coherent narrative equally well when we are faced with a choice (Kuczewski 1999). Intelligible and coherent narratives can even be crafted around fundamental changes that one strongly repudiates, or that are caused by oppressive social structures (Schechtman 2003). In short, a narrative picture of identity as the basis for making high-stakes decisions does not provide us with an indication of which changes will or could be incorporated seamlessly into the patient's narrative, and which might be experienced as troubling and disruptive to one's identity.[5]

Understanding psychological connectedness as the underlying basis upon which these decisions are made, on the other hand, does give us guidance in making these decisions, and is centred around what is most important to the patient. This picture suggests that interventions might be regarded as problematic in this way if they disrupt our settled, stable values, beliefs and other psychological elements *and* cannot be endorsed on the basis of our existing subjective motivational set. It should not be thought that psychological connectedness requires a static personality, so that any changes should be seen as problematic. A patient, for example, might value spontaneity, but have trouble acting in a spontaneous manner. A personality altering intervention that (as a foreseeable side-effect) increases spontaneity is likely to be welcomed by this patient as in line with his values. Because the personality change will be connected to the existing values of the patient in this way, the personality change will be seen as having, and will continue to have, high instrumental and representative value for the patient—the choice will allow the patient to pursue what he values, and to better express this value through his conduct. Such a change is likely to be regarded by the patient as increasing his authenticity—as bringing his conduct more into line with what he values and desires. If the patient thinks spontaneity is a character flaw, on the other hand, he might experience such a change as problematic, unwanted, and disruptive to his identity (see Nyholm and O'Neill 2016).

If we recognize psychological connectedness as a basis for making valued and valuable decisions, we understand the fear of 'not surviving' a personality altering procedure as a fear that the deeply held values, beliefs and desires that form the core of a patient's personality will be replaced by others, and will lose the ability to motivate him (see Schechtman 2003). Some degree of personality change is compatible with this picture; as long as changes are concordant with or endorsable on the basis of previous traits, establishing connections that stretch through time, psychological connectedness is retained. Such a picture gives us a basis for making these decisions in a way that captures what is valuable, and what is of concern to the patient, and thus gives us guidance concerning what we need to communicate to the patient, and how to understand and address the fear of not surviving personality change.

Establishing the importance of psychological connectedness as an underlying basis for making decisions of this type does not require a metaphysical commitment—like Jecker and Ko, I aim here to capture what concerns the patient has when he talks about survival, rather than making a metaphysical judgment about when we should say the patient has or has not survived. Thinking about this type of psychological connectedness when making these sorts of decisions can give us an indication of what consequences are likely to welcomed by, and experienced as valuable to, the

[5]Michael Kühler makes a similar point about narrative identity in the context of surrogate decision making in his paper in this volume.

patient, and which are more likely to be repudiated. I believe that some accounts of narrative identity might be compatible with recognizing the importance of the type of psychological connectedness I am emphasizing here (see e.g. Nyholm and O'Neill 2016), just as accepting biological continuity as a metaphysical theory of personal identity can be compatible with recognising the importance of psychological connectedness when it comes to issues of practical and ethical importance. But insofar as either approach does not make room for the importance of psychological connectedness as an important source of value for long-term, high-stakes decision making, they are unable to capture what makes decisions valuable to the patient, and what is of concern to them when making these decisions.

3.5 Conclusion

Part of the source of value of making one's own decisions is instrumental—it provides a way of achieving our goals or ends. Another source of value is representative—we want to express or represent what we value through our choices. Our choices only maintain this value over time if we remain psychologically connected to our former selves in certain ways—if our aims or values change, then the choice may no longer be valuable in these ways. This may make future-directed, self-regarding decisions less valuable in circumstances in which drastic personality change is imminent, and may weaken the moral authority of future-directed choice in these cases. This particularly applies to advance directives, where we are making a choice purely on behalf of our future self. There may be other good grounds upon which the authority of advance directives can be upheld, but we cannot identify the issues at stake here without delving into the sources of value of advance directives, and understanding the role that psychological connectedness has to play in this. Furthermore, reference to psychological connectedness allows us to understand what is of concern to the patient when she is facing potentially personality-altering interventions. It provides a picture underlying the patient's decisions that captures what she values—retaining certain central values and traits, while allowing for change that connects with existing values and priorities. Without underlying psychological connectedness, there is no sufficient basis upon which decisions can be made, and we cannot capture what the patient cares about in making decisions, distinguishing between changes that the patient is likely to welcome, and changes that she might find problematic or feel alienated from. To summarize; the patient values and aims for continued psychological connectedness in making decisions, and when these connections are severed, choices become less valuable, and their authority may be diminished. There is practical importance in exploring the interrelation between psychological connectedness and choice, with clear implications concerning how to understand and value self-regarding decisions in medical ethics, which comes apart from an abstract metaphysical debate about personal identity.

References

Baylis, Françoise. 2009. "I am who I am": On the Perceived Threats to Personal Identity from Deep Brain Stimulation. *Neuroethics* 6 (3): 513–526.

Blustein, Jeffrey. 1999. Choosing for Others as Continuing a Life Story: The Problem of Personal Identity Revisited. *Journal of Law, Medicine and Ethics* 27 (1): 20–31.

Buchanan, Allen. 1988. Advance Directives and the Personal Identity Problem. *Philosophy and Public Affairs* 17 (4): 277–302.

Buchanan, Allen, and Dan Brock. 1989. *Deciding for Others: The Ethics of Surrogate Decision Making*. Cambridge: Cambridge University Press.

Brock, Dan. 1991. Decision making competence and risk. *Bioethics* 5 (2): 105–112.

Darwall, Stephen. 2006. The Value of Autonomy and Autonomy of the Will. *Ethics* 116 (2): 263–284.

DeGrazia, David. 1999. Advance Directives, Euthanasia, and the Someone Else Problem. *Bioethics* 13 (5): 373–391.

Dresser, Rebecca. 2006. Dworkin on Dementia: Elegant Theory, Questionable Policy. In *Bioethics: An Anthology*, 2nd ed., ed. Peter Singer and Helga Kuhse, 65–73. Malden: Blackwell.

Duus-Otterström, G.öran. 2011. Freedom of the Will and the Value of Choice. *Social Theory and Practice* 37 (2): 256–284.

Dworkin, Gerald. 1994. Markets and Morals: The Case for Organ Sales. In *Morality, Harm and the Law*, ed. Gerald Dworkin, 155–161. Boulder: Westview Press.

Faden, Ruth, and Tom Beauchamp. 1986. *A History and Theory of Informed Consent*. Oxford: Oxford University Press.

Feinberg, Joel. 1986. *The Moral Limits of the Criminal Law Volume 3: Harm to Self*. New York: Oxford University Press.

Gillon, Raanan. 2003. Ethics Needs Principles—Four Can Encompass the Rest—and Respect for Autonomy Should be "First Among Equals." *Journal of Medical Ethics* 29 (5): 307–312.

Glover, Jonathan. 1977. *Causing Death and Saving Lives*. London: Penguin Books.

Grisso, Thomas, and Paul Appelbaum. 1998. *Assessing Competence to Consent to Treatment: A Guide for Physicians and Other Health Professionals*. New York: Oxford University Press.

Jecker, Nancy, and Andrew Ko. 2017. Is that the Same Person? The Problem of Identity Following Neurosurgical Procedures. *AJOB Neuroscience* 8 (3): 2–29.

Kuczewski, Mark. 1999. Commentary: Narrative Views of Personal Identity and Substituted Judgment in Surrogate Decision Making. *Journal of Law, Medicine and Ethics* 27 (1): 32–36.

McInerney, Peter. 1990. Does a Fetus Already Have a Future-Like-Ours? *Journal of Philosophy* 87 (5): 264–268.

Müller, Sabine, Merlin Bittlinger, and Henrik Walter. 2017. Threats to Neurosurgical Patients Posed by the Personal Identity Debate. *Neuroethics* 10 (2): 299–310.

Nyholm, Sven. 2018. Is the Personal Identity Debate a "Threat" to Neurosurgical Patients? A Reply to Müller et al. *Neuroethics* 11 (2): 229–235.

Nyholm, Sven, and Elizabeth O'Neill. 2016. Deep Brain Stimulation, Continuity Over Time, and the True Self. *Cambridge Quarterly of Healthcare Ethics* 25 (4): 647–658.

Oderberg, David. 1997. Modal Properties, Moral Status and Identity. *Philosophy and Public Affairs* 26 (3): 259–298.

Olson, Eric. 1997. *The Human Animal: Personal Identity without Psychology*. Oxford: Oxford University Press.

Parfit, Derek. 1984. *Reasons and Persons*. Oxford: Oxford University Press.

Scanlon, T.M. 1998. *What We Owe to Each Other*. Cambridge: Harvard University Press.

Schechtman, Marya. 1996. *The Constitution of Selves*. Ithaca: Cornell University Press.

Schechtman, Marya. 2003. Empathic Access: The Missing Ingredient in Personal Identity. In *Personal Identity*, ed. Raymond Martin and John Barresi, 238–259. Oxford: Blackwell.

Schechtman, Marya. 2010. Personhood and the Practical. *Theoretical Medicine and Bioethics* 31 (4): 271–283.

Taylor, James. 2009. *Practical Autonomy and Bioethics*. New York: Routledge.
Varelius, Jukka. 2006. The Value of Autonomy in Medical Ethics. *Medicine, Health Care and Philosophy* 9 (3): 377–388.
Warren, Mary Anne. 1977. Do Potential People Have Moral Rights? *Canadian Journal of Philosophy* 7 (2): 275–289.
Williams, Bernard. 1957. Personal Identity and Individuation. *Proceedings of the Aristotelian Society* 57: 229–252.
Witt, Karsten. 2017. Identity Change and Informed Consent. *Journal of Medical Ethics* 43 (6): 384–390.

Lucie White is an assistant professor at the Institute of Philosophy at Leibniz University Hannover, Germany. Her research interests include personal autonomy and applied ethics. Her autonomy- and personal-identity-focused publications include "The Need for Authenticity-Based Autonomy in Medical Ethics," in: *HEC Forum* 30(3), 2018, 191–209; "Personal Identity and Patient-Centered Medical Decision Making," in: *AJOB Neuroscience* 8(3), 2017, 194–195; "How Autonomy Can Legitimate Beneficial Coercion," in: J. Gather, T. Henking, A. Nossek, and J. Vollmann (eds.): *Beneficial Coercion in Psychiatry: Foundations and Challenges*, Münster: Mentis, 2017, 85–99; and "Understanding the Relationship Between Autonomy and Informed Consent: A Response to Taylor," in: *Journal of Value Inquiry* 47(4), 2013, 483–491.

Chapter 4
Personal Identity and the Moral Authority of Advance Directives

Karsten Witt

Abstract Many people believe that we formulate advance directives in order to care for our own future and that our directives have moral authority just when this is competently done. By implying that the signer of the directive cannot survive serious mental impairment, popular philosophical thinking about personal identity threatens to undermine this pair of convictions. Some philosophers react by denying the second conviction. According to them, advance directives should be followed even if the signer and the patient are two. The paper discusses three justifications for this claim and finds them all lacking. The ethics of advance directives cannot ignore the metaphysics of the person.

4.1 Introduction

Many people inside and outside of philosophy believe that advance directives for dementia and persistent vegetative state have moral authority. Normally, physicians should do what the directives say. However, it is unclear why this is. Why should physicians implement them? What does their authority rest upon? Many bioethicists believe that part of the answer has to do with personal identity. Advance directives, they say, have moral authority because the signer provides for her *own* future.

Some observers are not convinced. According to them, the moral force of advance directives does not depend on personal identity. In certain cases physicians should implement them even when the signer and the patient are distinct.

This controversy is at the heart of the present chapter. To understand what the dispute is about, I need to clarify my use of the term 'personal identity'. As is generally acknowledged, it can mean different things (Olson 2017, §1). When I use 'personal identity' or 'our identity' in this chapter, I am exclusively referring to the *numerical identity through time* of people (or persons) like you and me.

K. Witt (✉)
Köln, Germany
e-mail: karsten.witt@hspv.nrw.de

M. Kühler and V. L. Mitrović (eds.), *Theories of the Self and Autonomy in Medical Ethics*, The International Library of Bioethics 83,
https://doi.org/10.1007/978-3-030-56703-3_4

Numerical identity is the relation everything necessarily bears to itself and only to itself. It is *numerical* in that it tells us how to count: If *a* and *b* are identical, they are one; if *a* and *b* are distinct, they are two. It holds *through time*, if it holds between things existing at different times, between you and a certain boy or girl, say.

Most philosophers dealing with our identity through time focus on the *persistence question*:

> Under what conditions is a person existing at one time numerically identical with something existing at another time?

If we know the answer to this question, we can figure out which past or future being we are, when we started to exist, and when our existence necessarily ends. We can know, for instance, whether we once were tiny embryos or whether we can survive senile dementia or an irreversible loss of consciousness.

This chapter is about the numerical identity through time of the individuals involved in the process of setting up and implementing an advance directive: the person who signs it and the cognitively impaired patient, whose medical fate may be determined by it. I shall call them the *signer* and the *patient*. They are intimately related. They share one biological life and perhaps also one 'body'. (Whether they do depends, among other things, on what it means to have the same body, something I am unsure of; see Olson 1997, pp. 142–153.) However, in spite of their close links, their identity is sometimes disputed. As I will explain in §3, many philosophers explicitly or implicitly believe that in some cases involving advance directives the signer and the patient are two.

Although I will exclusively deal with the sense of 'personal identity' at issue in the persistence question, my focus is not on the persistence question itself. That is, I will not argue for or against a certain account of personal identity, be it biological, psychological, or what have you. To contribute to that debate is not my aim here. My interest is in a different question of personal identity. It asks for personal identity's role in the ethics of advance directives:

> Does the moral authority of advance directives depend on the numerical identity of signer and patient?

This *relevance question* will preoccupy me in what follows. I answer in the affirmative: The moral authority of advance directives does require the identity of the signer and the patient. Living wills can be morally authoritative only if the signer and the patient are one. This is my *dependence claim*. Various bioethicists argue against it. They say that living wills can retain their moral authority even if the signer and the patient are two. My aim in this chapter is to refute some of their arguments.

4.2 Three Preliminary Remarks

First, much of the ensuing discussion will refer to the ethics of advance directives in dementia and persistent vegetative state. This is due to the fact that the dependence claim has been disputed almost exclusively in these contexts. But this in no way limits my conclusions. If my argument is sound, personal identity is important for the validity of all sorts of advance directives, not only of advance directives in dementia and persistent vegetative state.

Second, if the dependence claim is true, the moral force of advance directives depends on personal identity. But personal identity alone cannot ensure their moral force. Other conditions must hold as well: When signing the directive, the signer must be competent and sufficiently informed. She must act voluntarily. And when she changes her mind due to new treatment options or new values, the directive must be updated. Some bioethicists argue that these additional conditions are hard or impossible to satisfy and that this casts doubt on the widespread use of advance directives (Buchanan and Brock 1990, pp. 103–107; Dresser 1995, p. 35; Vogelstein 2016). But these concerns are not my issue here. My claim is that even if they can be laid to rest, advance directives lack moral authority whenever signer and patient are two.

Third, that advance directives have 'moral authority' is often taken to imply that they have absolute moral weight: physicians must implement them in all relevant circumstances. Some bioethicists disagree. They point out that advance directives can sometimes be trumped by other moral considerations. Especially in cases involving the 'happy demented', severely demented patients apparently enjoying their lives, advance directives may be outweighed by considerations to do with the demented patient's current quality of life or with other morally relevant factors like, perhaps, the attitudes and emotions of the medical team or of the patient's caregivers.[1] That may well be correct. Therefore, when I write that advance directives have 'moral authority' or 'moral force', I have in mind the more modest thought that they have at least *considerable* moral weight. They give physicians a strong *pro-tanto* reason, but perhaps not an overriding or tie-breaking reason in every single case, for treating certain patients in a certain way. Personal identity is necessary for our living wills to have moral authority in this more modest sense.

4.3 The Trouble with the Dependence Claim

At first sight the truth of the dependence claim that our identity is necessary for the moral authority of advance directives is indisputable. When someone competently declares that a certain patient should under such-and-such future circumstances receive such-and-such medical treatment or, more frequently, that she should not receive a given treatment, it appears to be an all-important question whether she *is* that patient—whether she decides about her own or someone else's medical fate. According to conventional bioethical wisdom as well as common morality, deciding for others is much more problematic, and in need of a different kind of justification,

[1] See Dresser (1995, p. 37) and Birnbacher (2016, p. 290). I thank Susanne Boshammer for drawing my attention to this point.

than deciding for oneself. (You may, for instance, legitimately harm yourself, but not others.) And whether your decision is self-regarding or other-regarding turns on personal identity. It is self-regarding just when all people affected by it are numerically identical with you, and other-regarding just when it affects people numerically different from you.

More specifically, most bioethicists think that *self-determination* is the ground of advance directives' moral authority (e.g. Dworkin 1993; Quante 1999; DeGrazia 2005; Davis 2014). They should be followed because they 'extend' or 'prolong' the signer's autonomy to times of incompetence. But self-determination presupposes numerical identity. For an act to be *self*-determined, the determining agent and the agent who is determined must be one. Therefore, if bioethicists are right about the ground of advance directives' moral authority, personal identity is of crucial importance for the directives' practical bearing.

So there is strong prima-facie support for the dependence claim. However, the claim leads into rough waters when combined with one of the most popular answers to the persistence question. According to that answer, we cannot survive severe dementia and other conditions involving advanced or complete cognitive decay. We die, or anyway cease to exist, before we can become severely demented or permanently unconscious. This is because our continued existence presupposes a well-integrated mental life. For us to exist from one time to another there must be a certain number of psychological connections between the two times (or between the first time and an intermediate time, and between the intermediate time and the second time): connections like experiences and their memories, intentions and later acts in which the intentions are carried out, and persisting beliefs, desires, skills, and personality features. Since there are at best only few such connections between us and the severely demented or permanently comatose, we cannot be them, the reasoning goes.

Among bioethicists Derek Parfit is the most famous representative of such a view.[2] He writes:

> For X and Y to be the same person, there must be over every day enough direct psychological connections....[T]here is enough connectedness if the number of direct connections, over any day, is at least half the number that hold, over every day, in the lives of nearly every actual person. (Parfit 1984, p. 206)

A person persists for a day, Parfit writes, only if she retains at least half the number of 'direct psychological connections', connections of the sort described above, that we typically have over a day when we are people. Severely demented or permanently unconscious patients fall below this minimum. They are not connected with anyone in the way Parfit says is necessary for us to survive. So we cannot become such patients. We cannot become severely demented or fall into a persistent vegetative state. We cease to exist before that happens.

[2]Similar views (or views with similar consequences) are held by Baker (2000, p. 11), Shoemaker (2011, p. 360), and Lowe (2015, p. 146). Moreover, there is reason to believe that most contemporary adherents of the psychological view of personal identity are implicitly committed to something like the Parfitian position; see Olson (1997, pp. 22–27) and Witt and Olson (2020, § 2).

It is easy to see how the combination of a Parfitian answer to the persistence question and my dependence claim leads friends of advance directives into trouble. Severe dementia and persistent vegetative state involve a high degree of mental decay. A Parfitian account of our persistence excludes our existence in such states. We cannot become severely demented or permanently comatose because we cease to exist before that happens. Given the dependence claim that our advance directives' moral authority depends on our survival in such circumstances, they have no moral authority.

Here is the reasoning in a nutshell:

(1) In conditions involving far-reaching mental decay the signer and the patient are numerically distinct.
(2) If the signer and the patient are numerically distinct, advance directives have no moral authority.
(3) Thus, in conditions involving far-reaching mental decay advance directives have no moral authority.

This is the *identity problem*. Most bioethicists want to solve it by embracing a non-Parfitian answer to the persistence question like the 'embodied-mind view' (McMahan 2002) or the 'biological view' (Olson 1997), thus rejecting its first premise. But some are drawn to a Parfitian answer to the persistence question. This is why they accept the first premise.[3] But then the only solution open to them is the rejection of premise 2. And this is what they do. They claim that advance directives have moral authority even if signer and patient are two. In other words, they go against my dependence claim.

4.4 Avoiding Confusion

In §§ 6–8 I will consider and refute three solutions to the identity problem that deny the dependence claim. But before I come to this, I need to forestall a possible confusion and point to a general difficulty. I start with the confusion.

It is easy to mistake the dependence claim for a different though similarly sounding assertion: the claim that for an advance directive to have moral authority the signer and the patient must be the same person. This *same-person claim* is deceptively similar to the dependence claim. If the signer and the patient are the same person, are they not numerically identical? And, conversely, if they are not the same person, are they not numerically distinct? But, as I will argue now, it is vital for most bioethicists to keep the two claims apart. Their beliefs about our persistence, personhood, and the same-person relation allow for the possibility that the signer is the patient although they are not the same person. More precisely, they allow for the possibility that the signer, as she exists at t (the time of signing the directive), is numerically identical

[3]Perhaps they needn't, but this is another story; see Witt (2020).

with the patient, as she exists at t^* (the time when the directive is thought to apply), although the signer is, at t, not the same person as the patient at t^*.

Some bioethicists say that the signer is the patient, although, at t^*, the patient is not a person. They accept a 'Lockean' account of personhood according to which severely demented human beings or human beings in a persistent vegetative state are not persons (or people). So if, at t^*, the patient is severely demented or in a persistent vegetative state, she is not a person and *a fortiori* not the same person as anyone. They combine this with an answer to the persistence question which entails that we can survive the loss of personhood and become non-persons: because we persist in virtue of either brute-physical continuity (e.g. DeGrazia 2005) or 'weak' psychological continuity (McMahan 2002). This leads them to say that the signer is the patient although they are not the same person.

Other bioethicists come to the same conclusion, albeit on a different path. They claim that the signer is the patient, that both and that both are persons (i.e. they hold a 'non-Lockean' account of personhood), and that the signer is a different person from the patient. According to them each of us can in principle become a different person. We can be a certain person at one time and a different person at another. And one way to become a different person is to become severely demented or fall into a persistent vegetative state. This is roughly in line with everyday thought and talk. We often say about someone that 'she is not the person (or woman) she used to be' or that 'she has become a different person'. And we sometimes say this about our demented friends or relatives.

If the signer and the patient are numerically identical but not the same person in one of these ways, the dependence claim and the same-person claim have opposing implications for the moral authority of advance directives. While their bindingness is not in question according to the dependence claim (since signer and patient are one), it is undermined by the same-person claim (since signer and patient are not the same person).

Of course, one *can* hold beliefs about our persistence, personhood, and the same-person relation from which it follows that it is impossible for the signer and the patient to be numerically identical without being the same person. But few bioethicists do.[4] All others should carefully distinguish between the dependence claim and the same-person claim.

Not all parties to the debate seem to make the distinction. Rebecca Dresser's seminal discussion of the identity problem is perhaps most problematic in this regard. She adopts Parfit's answer to the persistence question and criticizes the use of advance directives in cases of severe mental disability in light of it. In such situations, she says, the signer is 'no longer in existence', while the patient is a 'new person' (Dresser 1986, p. 381). She adds that in such a case.

> there is no particular reason why the past preferences, as opposed to [the preferences of] any other person, should determine the present person's fate. (380f.; see also Dresser and Robertson 1989, p. 236; Dresser 1995, p. 34)

---

[4]Birnbacher might be an exception; see Birnbacher (2016, p. 286).

While this bit of reasoning seems to be based on the dependence claim, her rhetorical question,

> Why should a patient *who is now* a different person be burdened by a treatment decision consistent with the former person's preferences? (381, my emphasis)

and remarks such as

> Incompetent patients ... can receive no present benefit from treatment decisions in accord with *their* former preferences (ibid., my emphasis)

seem to trade on the same-person claim that for advance directives to have moral authority the signer and the patient must be the same person—and not (only) one and the same thing, as the dependence claim would have it. In light of such statements it is impossible for me to say whether Dresser bases her rejection of advance directives on the dependence claim or on the same-person claim (Recent discussions of the identity problem that mirror these unclarities are Davis 2014, 546f.; DeMarco and LiPuma 2016, and Müller et al. 2017, 303f.).

I do not know why the dependence claim and the same-person claim get mixed up so often. It might have to do with well-known ambiguities surrounding the phrase 'personal identity' and the problems going along with them (see again Olson 2017, §1). Or it might be because Parfit's answer to the persistence question and his 'reductionist' view of human people are so difficult to grasp. But whatever the correct explanation for the muddle ultimately is, bioethicists working on advance directives must avoid it. They must make clear whether they are dealing with the dependence claim or the same-person claim and then stick to their guns.

4.5 A General Difficulty

I am here dealing exclusively with the dependence claim and solutions to the identity problem that reject it. These solutions face a general difficulty. Since they accept that the signer and the patient are numerically different, they cannot ground the moral authority of advance directives on the patient's informed consent. Informed consent presupposes the consenting patient's competence (Beauchamp and Childress 2013, p. 124), but in the cases we are considering the patient is *in*competent and the competent signer is not the patient. So whatever treatment advance directives envisage for severely demented patients or patients in a persistent vegetative state, it cannot be legitimized by the patients' informed consent. The chief justificatory source of modern bioethics is unavailable for all solutions to the identity problem I am going to discuss. This is awkward, since advance directives typically contain refusals of life-sustaining treatment. If they are implemented, the patient dies. This is already difficult for physicians when she *has* given her informed consent. It is bound to be much more delicate when the life-shortening decision has been made by others (I revisit these issues in §8.).

4.6 The Network View

The first attempt to solve the identity problem by abandoning the dependence claim is put forth by Søren Holm. Based on Parfit's account of our identity he concludes that in cases of severe dementia signer and patient are 'two distinct persons'. But instead of deducing from this the invalidity of advance directives, he continues:

> It does, however, seem strange if our relations to demented persons should be governed solely by this piece of sophisticated philosophical reasoning about the ontological status of persons. Real persons are embedded in social networks and occupy social roles. My maternal grandfather kept on being my maternal grandfather, even at a time when his dementia had developed so far that he could no longer recognise me as his grandson. (Holm 2001, p. 157)

As Holm sees it, that the signer and the patient are numerically distinct is a sophisticated truth, but it should not govern our dealings with severely demented patients.[5] What *should* govern them is the fact that the patients are 'embedded in social networks and occupy social roles'. His demented grandfather is a ready example: He was Holm's grandfather when healthy, and he remains to be his grandfather when in the last stages of Alzheimer's. I do not think that Holm wants to say, absurdly, that grandfathers can survive severe dementia while people cannot. He makes no claim about our persistence (Again, he accepts Parfit's account.). His proposal refers to the relevance question. As I understand it, it says that what is necessary for the moral force of his grandfather's advance directive is not whether the old man still exists, but whether his family members and friends *think* that he still exists and treat a certain nursing home resident accordingly. Likewise, what is necessary for the moral bindingness of my advance directive is not whether I am numerically identical with a certain future patient, but whether that patient.

> is visited by my friends, belongs to my family, is cared for by my doctor, and even enjoys access to my bank account and all my worldly possessions. (Jecker 2016, p. 202)

In a word: what is necessary for the moral bindingness of my advance directive is not whether the patient is me but whether the people in my social network *believe* that he is me (see also Jox 2006, p. 87, Harvey 2006, p. 55, and Jecker 2016, p. 209).

This is the *network view*. Its adherents do not seem to be overly confident about it. It is usually presented as one argument among others in joint support of the claim that the numerical identity between signer and patient is unnecessary for advance directives to apply. But even the attenuated claim is wrong. The network view cannot carry any justificatory weight. It cannot be part of a joint endeavour to secure the moral authority of advance directives in the absence of numerical identity. Closer scrutiny shows it to be untenable.

One problem is its vagueness. Friends of the network view only gesture at the set of people and social roles that is crucial for their proposal to work. Perhaps the roles are those defined by civil law (as Merkel 2004, p. 304, n. 7, proposes); but

---

[5] In the quoted passage Holm writes that the lack of identity should not 'solely' govern the relations between signer and patient. But I have been unable to find what the limited role of personal identity that is implied by this remark might be.

who is, and who is not, a member of the patient's 'social network' is anyone's guess. Moreover, it is assumed that the people in the network always agree on matters of personal identity. But whether they do is by no means clear. People may have diverging opinions on whether the patient and the signer are one or two, and hold their respective beliefs with varying degrees of confidence. The network view offers not even the beginning of an answer to such disputes. It does not tell us what to do when, in a given case, the patient's social network is divided on the question of her survival or when there is widespread doubt about what to say. Do advance directives in such constellations have any chance of being morally binding? Is there a threshold of approval (or certainty) that has to be met? Without answers to these questions the network view works, if at all, only in a very limited range of situations.

Perhaps it can be sharpened to avoid these difficulties. But others remain. One is that the proposal has repugnant consequences when generalized. It tells us that, when the moral authority of advance directives is at stake, we should not rely on 'sophisticated philosophical reasoning about the ontological status of persons' but on people's beliefs. But now consider that frightfully many people hold, typically on poor evidence, that there is no such thing as climate change or at least no climate change caused by human doings, or that ingesting the powder made from rhinoceros horns increases male libido, or that homosexuality is a disease. Should we conclude from this that the moral authority of the recommendations of the IPCC, the WWF, or the LGBT movements are relativized or undermined? If we think that the recommendations' moral force depends (in a suitably sharpened way) on people's beliefs, the answer should be Yes. But I expect most readers will emphatically answer No. What matters are not people's beliefs but the insights of climatology, biology, and medicine. Why then should things be different with the moral authority of advance directives? Why should they be based on people's beliefs and not on the insights from 'the ontology of persons'?

More generally, the network view's underlying idea, that the force of moral recommendations depends on what people happen to believe, will almost inevitably have repugnant consequences when applied to other areas. The reason is that it does not discriminate between beliefs of varying epistemic strength: it counts beliefs nourished by prejudice, hate, or echo-chamber conspiracies just as much as beliefs sustained by the science of the day. Obviously, this cannot be true. And I am sure that proponents of the network view will reject any such generalization. The difficulty they face is to consistently reject it while insisting that false beliefs about the persistence of Holm's grandfather *are* authority-conferring.

Note that even people contemptuous of contemporary science will, on consideration, reject the generalized network view. Like us, they will insist that it is not *beliefs* about climate change, the aphrodisiac power of rhino keratin, or about the nature of homosexuality that should determine whether the recommendations of the IPCC, the WWF, or the LGBT movements have moral authority, but the facts: whether the climate changes, rhino horns are libido-enhancing, or homosexuality can literally be cured. The dispute between them and us is not on the authority-conferring status of beliefs, but on what counts as a reliable source of knowledge. They will reject any proposal that makes the authority of their recommendations dependent on what we believe. (That the proposal comes from philosophers is likely to raise their suspicion.)

This brings out another serious problem for the network view. Most laymen not only believe that we can survive severe dementia; they also believe that this ability is part of the reason why living wills are morally authoritative. That the moral authority of advance directives depends on our persistence and not our beliefs about our persistence is as intuitively compelling as the claim that the moral authority of the advice to industrial countries to reduce their carbon emissions depends on the climate and not on our beliefs about the climate. The *dependence belief*, that the moral force of advance directives depends on personal identity, thus appears to have no less intuitive credit as the *persistence belief*, that we can continue to exist through all stages of dementia and coma. But the network view treats the two beliefs in markedly different ways: while it highlights the latter, it ignores the former.

This is problematic. Return once more to the persistence belief and assume, as Holm and others do, that it prevails among the people in the signer's social network. Friends of the network view say it is false, but matters nevertheless: it ensures (or contributes to ensure) the moral authority of the signer's advance directive. This is incompatible with the dependence belief which, as we might plausibly assume, also prevails in the patient's network. It says that without our persistence the moral authority of advance directives is null and void. Unsurprisingly, friends of the network view say that it is false too. (It is their central claim.) But, given that false persistence beliefs matter for the moral authority of advance directives as long as they prevail in the patient's social network, shouldn't false dependence beliefs matter as well? Shouldn't friends of the network view take both beliefs at face value and integrate them equally into their account?

The answer is that they cannot do this. Since the persistence belief is false and the dependence belief implies that false persistence beliefs don't count for the moral authority of advance directives, Holm and his followers cannot accept both the persistence belief and the dependence belief. Taking the dependence belief seriously points them to our persistence, not our false beliefs about our persistence, while taking the persistence belief seriously commits them to the exactly the opposite. Adherents of the network view therefore cannot base their solution of the identity problem on the principle that decisions about severely demented patients should be guided by *all* relevant beliefs in the patient's social network. But it is hard to see how a more limited principle, one that excludes the dependence belief but not the persistence belief, could be justified. The network view is either incoherent or arbitrary.

4.7 The Continuer View

In another influential reply to the identity problem, Jeffrey Blustein argues that the numerical distinctness of signer and patient 'does not settle the issue of moral authority'. What settles it is whether by implementing advance directives people act as '*continuers* of the life-stories of those who have lost narrative capacity' (Blustein

1999, p. 21). The moral authority of advance directives requires that the signer tells (or would tell) a suitable story of her life that identifies her with the patient and that the implementation of the advance directive is an apt continuation of that story. Numerical identity is not needed.[6]

Note that Blustein's formulation is slightly misleading. Saying that his view concerns patients who have 'lost' narrative capacity, suggests that these patients *were* once able to tell stories. But then they were also able to sign, and actually did sign, an advance directive, so that the identity problem does not arise. However, given Blustein's repeated commitment to the view that in severe dementia or permanent unconsciousness signer and patient might be two (on pp. 20, 22, 27, 30), this cannot be what he has in mind.

What is distinctive about this *continuer view* is its emphasis on the signer. While the network view focusses on the persistence beliefs held by people in the patient's environment, the continuer view starts out with the signer's own persistence belief. Does she think she can become severely demented or permanently unconscious? Does she think that severe dementia or irreversible coma can be part of the story of her life? If so, her caretakers have a strong *prima-facie* reason to act as continuers of her story and comply with her advance directive.

Or rather, it is the people who consider themselves to be her caretakers, who have the strong reason. Whether they really take care of *her*, or of a numerically distinct being does not matter according to the continuer view. What matters are the patient's story and its continuation by the implementation of the advance directive. The final chapter need not accord with the facts. It lends authority to the signer's directive even though she never becomes severely demented.

The continuer view needs constraints. If just any old story would qualify as authority-grounding and any continuation as apt, it would be a nonstarter. Not only could the patient tell several incompatible stories, making it hard for the caregiver to choose which one to follow up upon. The stories could also be intentionally made up or even born in an insane delusion. No one would want rest a life-shortening decision on them. Furthermore, if any decision whatever would count as an apt continuation of the patient's life-story, the caregiver could opt for medical interventions which are completely at odds with the patient's interests and values and yet continue her life-narrative. Again, no one wants this. Blustein therefore proposes two constraints: eligible life-stories must not be 'distorted by self-deception, or … out of touch with reality, or seriously out of synch with the view that others have of us' (1999, p. 23); and their continuation by caregivers should be 'faithful to who the patient was, as expressed in the organizing principles of her life' (ibid.). Call the first the *reality constraint*, the second the *condition of narrative fit*.[7]

[6]See also Kuczewski (1994, 42f.) and Jecker (2016, pp. 206–208); for critical discussion see DeGrazia (2005, pp. 183–186), Shoemaker (2009, pp. 180–183).

[7]Note once more how the second condition's formulation muddies the waters. In demanding that narrative fit requires continuers to be 'faithful to who the *patient* was' (my emphasis) the quoted passage suggests that the patient is the signer—something Blustein officially denies. Given his Parfitian assumptions about our persistence, the continuation of the signer's narrative can only be faithful to who the *signer* was.

The condition of narrative fit appears to be reasonably clear in the cases that interest me: In situations in which an advance directive exists, it should be easy for proxy decision-makers to know how to continue the story of the signer's life in an adequate way. But, then again, it may not be so easy. Even when a living will exists, narrative fit can be difficult to ascertain when it is unclear whether the signer has changed her mind after signing the document. Perhaps she excluded life-sustaining medical treatment in the directive, but since then occasionally expressed her approval of critical writings on non-treatment directives. Should the caregiver conclude from this that implementing the living will would no longer be faithful to the story of the signer's life? This is a difficult question. However, the difficulty is not unique to the continuer view. Any account that accords moral authority to advance directives has to deal with similar problems (§2). I therefore set them aside.

At any rate, the reality constraint is more troublesome for the continuer view. It says that eligible life-stories must not be 'out of touch with reality' (I take the other elements of the constraint mentioned above to be variations of this basic idea.) But this is what they clearly seem to be. Their final chapter is based on the author's assumption that she can survive severe dementia. But the assumption is false. Not *we* are the incompetent protagonists of our life-story's final chapter, but a numerically different being. This appears to violate the reality constraint.

The violation is inevitable. For life-stories to bridge the gap between signer and patient, their final chapter must contain a host of false claims about their protagonist. The claims could only be true if there were nothing to bridge; if the signer still existed. But then the identity problem would not arise in the first place. The continuer view, aspiring to be a solution to that problem, would be redundant. So the continuer view is worthy of consideration only if the stories it rests upon violate the reality constraint. It might be a candidate for solving the identity problem, but by its own standards is doomed to failure.

These dismal prospects originate in a construction error at the foundation of the continuer view. It is designed to a handle a situation of non-identity but contains a constraint that appears to require precisely the opposite. But perhaps the reality constraint is less stringent than I suppose (see Blustein 1999, p. 26). If so, the continuer view's proponents should be much more explicit about what they have in mind.

In any event, the reality constraint is not their only worry. Like the network view their justification of the moral force of advance directives is built on false persistence beliefs held by certain people. So it seems to be open to the same objection of arbitrarily favouring deeply held convictions about personal identity over deeply held convictions about the relevance of personal identity for the moral authority of advance directives. Can friends of the continuer view avoid this charge?

At first glance they can. They not only claim that the moral authority of advance directives depends on stories. They also assert that the stories which ensure the validity of advance directives are the things that constitute personal identity (Kuczewski 1994, p. 42; Blustein 1999, p. 21; Jecker 2016, p. 206). Their commitment to a 'narrativist' account of personal identity allows them to say that personal identity *is* necessary for the moral authority of advance directives. Our identity is our life-story. So it is necessary for our living wills to have moral force. Unlike the network view, then, the continuer view does not seem to treat basic intuitions in an arbitrary way. It takes at face value the belief that we can survive far-reaching or even complete mental devastation *and* the intuition that the moral authority of advance directives depends on personal identity.

This defence fails. It trades on a well-known ambiguity of the term 'personal identity'. The term can refer to our numerical identity through time. This is how I use it. But the friends of the continuer view, like most other narrativists about personal identity, use it to refer to a person's 'self-conception' (Blustein 1999, p. 23), or the set or bundle of 'beliefs, values, desires, and other psychological features [that] make someone the person she is' (Schechtman 1996, 1f.; see also Olson 2017, §1). This *characterization identity* is important in many practical deliberations. It is what most non-philosophers have in mind when they talk about 'identity': when they say that someone is having an 'identity crisis', for instance, or when certain medical interventions are claimed to have 'identity-changing' effects on patients. But it is markedly different from our numerical identity through time. What makes us the person we are at any one time and what ensures our existence over time are at best loosely related (Shoemaker 2006, p. 41). So even if the continuer view is entitled to say that the moral authority of advance directives depends on personal identity *in the characterization sense*, it is not entitled to say that it depends on our identity in the sense at issue in the persistence question. But it is 'personal identity' in this latter sense that figures in the dependence belief. Like the network view, the continuer view neglects it without further ado, while at the same time appreciating widespread views about what we can survive. In addition to home-made problems with constraints on narratives it is arbitrary and unprincipled in the same way, and for the same reasons, as the network view.

4.8 The Pragmatic View

One of my complaints against the network view and the continuer view is their apparent arbitrariness. They accept (what they consider) false beliefs about our persistence and reject (what they consider) false beliefs about our persistence's role in grounding the moral authority of advance directives without saying why. But perhaps there is a reason for us to accept widespread persistence beliefs or the life-stories incorporating them. Someone might argue as follows: Most people who sign an advance directive cherish the prospect of autonomously arranging what should happen to them in times of incompetence (Birnbacher 2016, p. 289). So even if bioethicists take the persistence belief to be false, they had better leave the valued prospect intact and act as if the directives had moral authority. People may err about their persistence, but the use of living wills is still justified by its overall good consequences. Advance directives are simply too beneficial to be abandoned.

This is the *pragmatic view*. We can find it in Helga Kuhse's discussion of the moral authority of advance directives. To solve the identity problem, she says, we should 'look for a justification of advance directives in the value they have for their authors'. More to the point, she writes:

> The advance refusal of life-sustaining treatment will satisfy the interests or preferences of the author of the directive to control what she sees as her future without thereby thwarting any of the demented patient's interests. (Kuhse 1999, p. 361)

DeMarco and LiPuma appear to agree:

> The issue is about respecting the interests of the now-competent person … Violating advance directives can make a profound difference to the living (DeMarco and LiPuma 2016, p. 682).[8]

Although it resembles the other two approaches discussed so far in its willingness to accept people's beliefs, the pragmatic view can avoid most of the objections they face. By appeals to the overall good its defenders seem to be able to explain why we should accept widespread persistence beliefs, but not dubious opinions on climate change, rhinoceros horns, or homosexuality; why it is legitimate for bioethicists to focus on people's persistence belief, but not on their belief that advance directives' moral authority depends upon our persistence; and why it does not matter that the stories told by signers and their continuers are false. In all these cases we promote the overall good. So the pragmatic view appears to have important advantages over the other views discussed so far. Nevertheless I doubt whether many bioethicists will welcome it as a solution to the identity problem.

First, like the other proposals it is incompatible with two of their central tenets: that ethically legitimate medical treatment presupposes the patient's informed consent and that medical care should be focussed exclusively on the patient. I begin with *informed consent*. As mentioned above, most advance directives for permanently incompetent patients contain refusals of life-sustaining treatment. If a non-treatment directive is implemented, the patient dies earlier than she would otherwise have. But if the signer is not the patient, the decision that is shortening the patient's life is not her own. It is the signer's. The patient has never competently agreed to her own premature death (§5).

That certain patients are left to die prematurely without their informed consent need not contradict the contemporary bioethical mainstream. Bioethicists are well aware that an incompetent patient's autonomous preferences are sometimes unknown. In such cases most of them say that physicians and caregivers should act as 'surrogate decision-makers' and do what is in the patient's best interest (Beauchamp and Childress 2013, pp. 228f.). But this is not how pragmatists argue. They place significant moral weight on the 'comfort and assurance of the competent signer' (Kuhse 1999, p. 358) and her 'interest … to control what she sees as her future' (361). Even if Kuhse writes that 'the interests of the surviving incompetent patient must also be taken into account' (358), she does not assign them special weight. They can thus be easily outweighed by the more sophisticated interests of the signer.

[8] See also Müller et al. (2017, 300f). Some philosophers claim that advance directives have moral force because they contain the signer's 'surviving interests' and because satisfying these interests improve the signer's life even after she is dead and gone; see Buchanan and Brock (1990, pp. 162–164) and Shoemaker (2010, 488f). The view is obviously related to the pragmatic approach, but deserves a discussion of its own.

The pragmatic view thus seems to be at odds not only with the principle of informed consent but also with the principle of *patient-centeredness*—that medical doctors should stand by their patients (Witt et al. 2017). Its defence of the moral authority of advance directives is more in line with utilitarianism than with the broadly deontological outlook shared by most bioethicists today. This makes it an unattractive solution of the identity problem to begin with. (The kind of utilitarianism that fits the pragmatic view might not even be attractive to utilitarians; see Furberg 2012, pp. 65–67.)

Second, it is unclear whether the pragmatic view succeeds on its own terms. It argues that advance directives have moral authority largely because their positive effects for the competent signer outweigh their negative effects for the patient. (As stated in the above quotation, Kuhse thinks that letting severely demented patients die is not 'thwarting any of [their] interests'. But this is bound to be controversial.) But the calculation elides too many interested parties. Family members might not want to let (what they take to be) their demented relative die. They might consider it a very bad thing not to help her when she is most vulnerable. The members of the medical team might think and feel similarly. Helping the patient is what, for many of them, their work is all about. They will almost certainly find it hard or even intolerable to implement non-treatment directives. (There is evidence that many of them do; see, e.g., Schöne-Seifert et al. 2016.) And, finally, we, as members of society, might 'hate living in a community' whose physicians have to deny some of their patients life-saving treatment (Dworkin 1993, 228f.). Including these interests in the pragmatic calculus might tip the scales against the moral authority of advance directives. The pragmatic view might turn out to be self-undermining.

Of course, people's attitudes are not carved in stone. Friends of the pragmatic view could teach family members, medical staff, and society at large to accept, or even to appreciate, treatment refusal grounded in advance directives as an expression of the patient's self-determination. In this way they could lessen the concerns of relatives, medics, and the society at large. But, given what they believe about our persistence, they would be teaching something false. Why not instead teach the truth? Why not try to enlighten people about their persistence conditions? This might well be difficult. But, if successful, it might ease people's concerns about dementia in its own way: people would come to see that *they* will never become severely demented. The illness' last stage will be none of their business. So it seems that once the pragmatic view adopts a *developmental perspective*, a perspective that takes into consideration the good or bad consequences of changing people's attitudes, it is at least an open question whether its proponents should confirm or deny the moral authority of advance directives: whether they should praise them as autonomy-enhancing or denounce them as offering relief for concerns nourished by illusions.

For that matter, isn't there an important moral difference between teaching people falsehoods and enlightening them? Don't philosophers have a moral obligation to promote what they think is true, even if spreading illusions proves more beneficial overall (Brock 1993; Lohmar 2017, pp. 313–317)? If they have the obligation, pragmatists taking the developmental perspective have an even stronger reason to stop defending advance directives.

My previous two criticisms add up to a third. The pragmatic view demands that incompetent patients be left to die for other people's comfort and assurance. To my mind this a bad, even frivolous reason for letting patients die. They may legitimately die for their *own* comfort: if they are tormented by anxiety and debilitating diseases, for instance, as is often the case in severe dementia; or if their future is mentally void due to the destruction of their higher brain as in persistent vegetative state. But can it ever be morally right that they die for the comfort and assurance of someone else?

According to a universal moral intuition it is outrageous to kill a person in order to use her organs to save several other persons' lives (Alexander and Moore 2016, §2.2). If so, killing a person for other people's comfort and assurance is surely even more repugnant. And so is letting her die when she could easily be saved through medical treatment. Now, many bioethicists working on advance directives tie personhood to sophisticated cognitive abilities like rationality and self-consciousness. According to them, severely demented human beings or human beings in a persistent vegetative state are not persons because they lack these abilities (§4). If they are right, the intuition that we must not let a person die for the comfort and assurance of other people offers no support in the cases at hand. We should rather ask whether it is morally permissible to let a human non-person die in order to provide comfort and assurance to a person. And I presume that most of us would again answer No. It appears impermissible to let an incompetent human being die for other people's relief. Many observers would not even countenance letting her die to save others' lives. (They would not do it if she were a newborn child.)

Proponents of the pragmatic view might reply that these judgements are distorted by what we normally associate with a human life: that it is (or will be) rich, evocative, and important; full of plans and meaning and depth. But the lives of the severely demented and those in a persistent vegetative state are not like this. They resemble the lives of most non-human animals (Kuhse 1999, 358f.). If pleasurable at all, they contain only experiential pleasures. Besides they are short. So even the happy demented don't lose much by dying prematurely. On the other hand there is considerably more to lose for the signer and (perhaps) for the people related to her. Without the possibility of advance directives the signer might have recurring worries about prolonged mental decay and lack of control; and her family and friends might spend much energy, time, and money on a patient they mistakenly take to be her. The premature death of the patient is an acceptable price to pay for the heightened comfort and assurance of these people.

I remain sceptical. There seems to me to be something profoundly wrong about letting highly vulnerable patients die, not because it is their autonomous wish or because it is in their own best interests, but solely for the sake of others. Furthermore, the good consequences that pragmatists think can justify the moral force of advance directives rest on false beliefs about our persistence: If people knew that neither they nor their loved ones can ever become severely demented or irreversibly comatose, and if they knew that the other people know, advance directives would offer neither comfort nor assurance. So they would not have moral authority on the pragmatic approach. And no incompetent patient would have to die solely for the sake of others.

The justification of the pragmatic approach's repugnant consequences is based upon prejudice and illusion. If anything, this should make it even more repugnant.

4.9 Final Remarks

The identity problem threatens the moral authority of advance directives. This chapter has been dedicated to three solutions that have been proposed in the literature. They say that advance directives have moral force even if signer and patient are numerically distinct. If they are right, bioethicists dealing with advance directives need not worry about personal identity. Whether they should be implemented is a purely practical question. Metaphysical considerations do not enter the picture.

I have found all three solutions lacking. They are vague, difficult to generalize, arbitrary, incompatible with cherished principles in contemporary bioethics, self-undermining, or replete with repugnant consequences. I cannot recommend them. It very much looks as if the identity problem cannot be solved without thinking about personal identity. The ethics of advance directives needs the metaphysics of the person.

References

Alexander, Larry and Moore, Michael. 2016. *Deontological Ethics*, ed. by E.N Zalta. Stanford Encyclopaedia of Philosophy.

Baker, Lynne Rudder. 2000. *Persons and Bodies: A Constitution View*. Cambridge: Cambridge University Press.

Beauchamp, Tom L., and James F. Childress. 2013. *Principles of Biomedical Ethics*. New York: Oxford University Press.

Birnbacher, Dieter. 2016. Patientenverfügungen und Advance Care Planning bei Demenz und anderen kognitiven Beeinträchtigungen. *Zeitschrift für Ethik in der Medizin* 28: 283–294.

Blustein, Jeffrey. 1999. Choosing for Others as Continuing a Life Story: The Problem of Personal Identity Revisited. *Journal of Law, Medicine and Ethics* 27: 20–31.

Brock, Dan W. 1993. Truth or Consequences: The Role of Philosophers in Policy-Making. In *Life and Death. Philosophical Essays in Biomedical Ethics* ed. by Dan W. Brock, 408–416. Cambridge: Cambridge University Press.

Buchanan, Allen E. and Dan W. Brock. 1990. *Deciding for Others. The Ethics of Surrogate Decision Making*. Cambridge: Cambridge University Press.

Davis, Dena S. 2014. Alzheimer Disease and Pre-Emptive Suicide. *Journal of Medical Ethics* 40: 543–549.

DeGrazia, David. 2005. *Human Identity and Bioethics*. Cambridge: Cambridge University Press.

DeMarco, Joseph P., and Samuel H. LiPuma. 2016. Dementia, Advance Directives, and Discontinuity of Personality. *Cambridge Quarterly of Healthcare Ethics* 35: 674–685.

Dresser, Rebecca. 1986. Life, Death, and Incompetent Patients. Conceptual Infirmitics and Hidden Values in the Law. *Arizona Law Review* 28: 373–405.

Dresser, Rebecca. 1995. Dworkin on Dementia. Elegant Theory, Questionable Policy. *The Hastings Center Report* 25 (5), 32–38.

Dresser, Rebecca, and John A. Robertson. 1989. Quality of Life and Non-Treatment Decisions for Incompetent Patients. A Critique of the Orthodox Approach. *Law, Medicine & Health Care* 1: 234–244.

Dworkin, Ronald M. 1993. *Life's Dominion. An Argument About Abortion, Euthanasia, and Individual Freedom.* New York: Knopf.

Furberg, Elisabeth. 2012. Advance Directives and Personal Identity: What is the Problem? *Journal of Medicine and Philosophy* 37: 60–73.

Harvey, Martin. 2006. Advance Directives and the Severely Demented. *Journal of Medicine and Philosophy* 31: 47–64.

Holm, Soren. 2001. Autonomy, Authenticity, or Best Interest: Everyday Decision-Making and Persons with Dementia. *Medicine, Health Care and Philosophy* 4: 153–159.

Jecker, Nancy. 2016. Advance Care Planning: What Gives Prior Wishes Normative Force? *Asian Bioethics Review* 8: 195–210.

Jox, Ralf. 2006. Der 'natürliche Wille' als Entscheidungskriterium: Rechtliche, handlungstheoretische und ethische Aspekte. In *Entscheidungen am Lebensende in der modernen Medizin: Ethik, Recht, Ökonomie und Klinik*, ed. Jan Schildmann, Uwe Fahr, and Jochen Vollmann, 73–90. Berlin: Lit Verlag.

Kuczewski, Mark G. 1994. Whose Will Is It, Anyway? A Discussion of Advance Directives, Personal Identity, and Consensus in Medical Ethics. *Bioethics* 8: 27–48.

Kuhse, Helga. 1999. Some Reflections on the Problem of Advance Directives, Personhood, and Personal Identity. *Kennedy Institute of Ethics journal* 9 (4): 347–364.

Lohmar, Achim. 2017. *Falsches moralisches Bewusstsein. Eine Kritik an der Idee der Menschenwürde.* Hamburg: Meiner.

Lowe, E. J. 2015. The Probable Simplicity of Personal Identity. In *Personal Identity* ed. by Georg Gasser and Matthias Stefan, 137–155. Cambridge: Cambridge University Press.

McMahan, Jeff. 2002. *The Ethics of Killing. Problems at the Margins of Life.* New York: Oxford University Press.

Merkel, Reinhard. 2004. Zur Frage der Verbindlichkeit von Patientenverfügungen. Eine notwendige Ergänzung der bisher in Deutschland geläufigen Argumente. *Zeitschrift für Ethik in der Medizin* 16: 298–307.

Müller, Sabine, Merlin Bittlinger, and Henrik Walter. 2017. Threats to Neurosurgical Patients Posed by the Personal Identity Debate. *Neuroethics* 10: 299–310.

Olson, Eric T. 1997. *The Human Animal. Personal Identity Without Psychology.* Oxford: Oxford University Press.

Olson, Eric T. 2017. Personal Identity. In *Stanford Encyclopaedia of Philosophy*, ed. by Edward N. Zalta.

Parfit, Derek. 1984. *Reasons and Persons.* Oxford: Clarendon Press.

Quante, Michael. 1999. Precedent Autonomy and Personal Identity. *Kennedy Institute of Ethics Journal* 9: 365–381.

Schechtman, Marya. 1996. *The Constitution of Selves.* Ithaca, London: Cornell University Press.

Schöne-Seifert, Bettina, Uerpmann, Anna Lena, Gerß, Joachim, Herr, Davi 2016. Advance (Meta-) Directives for Patients with Dementia Who Appear Content: Learning from a Nationwide Survey. *Journal of the American Medical Directors Association* 17 (4): 294–299.

Shoemaker, David. 2009. *Personal Identity and Ethics. A Brief Introduction.* Peterborough: Broadview Press.

Shoemaker, David. 2010. The Insignificance of Personal Identity for Bioethics. *Bioethics* 24 (9): 481–489.

Shoemaker, Sydney. 2006. Identity and Identities. *Daedalus* 135: 40–48.

Shoemaker, Sydney. 2011. On What We are. In *The Oxford Handbook of the Self*, ed. Shaun Gallagher, 352–371. Oxford: Oxford University Press.

Vogelstein, Eric T. 2016. Autonomy and the Moral Authority of Advance Directives. *Journal of Medicine and Philosophy* 41 (5): 500–520.

Witt, Karsten. 2020. In Defence of Advance Directives in Dementia. *Pacific Philosophical Quarterly* 101: 2–21.

Witt, Karsten, and Eric T. Olson. 2020. Against Person Essentialism. *Mind*. DOI 10.1093/mind/fzaa016.

Witt, Karsten, Johanne Stümpel, and Christiane Woopen. 2017. Caregiver Burden and the Medical Ethos. *Medicine, Health Care and Philosophy* 20: 383–391.

Karsten Witt is Professor for Ethics and Sociology at the University of Applied Sciences for Police and Public Administration, North Rhine-Westphalia. He is author of one book and several articles on ethics of economics, bioethics, and personal identity. His most relevant recent publications are "Narrative and characterization," in: *Erkenntnis* 85, 2020, 45–63; "In defense of advance directives in dementia," in: *Pacific Philosophical Quarterly* 101, 2019, 2–21; "Narrative and persistence," in: *Canadian Journal of Philosophy* 49, 2019, 419–434 (together with Eric T. Olson); "Demenz und personale Identität" (Dementia and personal identity), in: *Zeitschrift für Praktische Philosophie* 5, 2018, 153–180; and "Identity change and informed consent," in: *Journal of Medical Ethics* 43, 2017, 384–390. His contribution to the present volume is generously funded by the Deutsche Forschungsgemeinschaft [WI 4519/2-1].

Chapter 5
Narratives in Flux. Why Patients' Life Stories Do Not Provide Decisive Instructions in Cases of Surrogate Decision-Making

Michael Kühler

Abstract This paper critically discusses the role narrative accounts of the self might play in surrogate decision-making. Medical cases in which patients are not or no longer able to express their interests and also have not given a prior account of them present us with the question of how to respect their autonomy, i.e. how the patient would probably want to be treated, were he or she able to decide for him- or herself. In such situations it seems natural to refer to a notion of the person's identity or self, for based on this we might extrapolate how the person would decide in an authentic manner. In this regard, narrative theories of the self seem to be able to provide a reasonable answer. Knowing a person's life story seems to suggest a simple and convincing way to determine her authentic interests based on how her story, i.e. her narrative identity, would continue in a plausible and coherent way. However, I claim that narrative accounts of the self inherently fail to provide a simple and convincing answer in such cases. After distinguishing between mere expressive narrative accounts and constitutive narrative accounts, I argue that expressive narrative accounts do not present an independent analysis of the constitution of the self to begin with and are at best a helpful epistemic tool to find out what the patient's autonomous preference likely would be. Yet, even this only holds under the contestable assumption that the patient's self remains stable. Constitutive narrative accounts, on the other hand, while indeed presenting an independent account of the constitution of the self, either revert back to an expressive narrative account if the idea of one's free authorship of one's life story is rejected, or they fail to provide any reason for why one should choose a certain version of the patient's life story as the newly to be constructed one. Instead, I conclude that a person's life story remains in a constant state of flux, which is why also constitutive narrative accounts are unable to provide surrogate decision-makers with a well-justified answer which treatment option to choose in order to respect the patient's autonomy based on who he or she is.

M. Kühler (✉)
University of Münster, Münster, Germany
e-mail: m.c.kuhler@utwente.nl; michael.kuehler@uni-muenster.de

University of Twente, Enschede, The Netherlands

M. Kühler and V. L. Mitrović (eds.), *Theories of the Self and Autonomy in Medical Ethics*, The International Library of Bioethics 83,
https://doi.org/10.1007/978-3-030-56703-3_5

5.1 Introduction

Respect for autonomy is one of the core principles in modern ethics, as is, consequently, also respect for patients' autonomy in medical ethics.[1] There is no doubt that patients' autonomous decisions about how they (do not) want to be treated cannot simply be ignored.[2] In medical practice, this typically takes the form of the patient giving his or her informed consent to, or rejection of, the medical treatment in question. However, and aside from the major problem of defining what autonomy, and the respect thereof, is exactly, there are obviously situations in which patients are not, or no longer, able to make an autonomous decision and give their informed consent. For example, a patient might simply be unconscious or otherwise mentally incapacitated, or severely limited in his capacity to make autonomous decisions, be it only temporarily or permanently. In such cases, including at least some cases of palliative care and end-of-life decisions, and assuming that the patient has not provided an advance directive, decisions usually have to made by surrogates.[3]

When it comes to surrogate decision-making, i.e. when it comes to a situation in which others have to make a decision for the patient in his or her best interest, it seems only natural for surrogates to try to make a decision that the patient him- or herself would make in the situation, were he or she capable of deciding autonomously. After all, the issue at hand still concerns respecting the patient's autonomy, whereas surrogates now face the problem of finding out what the patient's autonomous decision would be.

Given that autonomy basically means self-government, i.e. the capacity to decide for *oneself* how to act, how to be treated, or, in general, how to live, surrogates obviously have to refer to the idea of the patient's *self*[4] in order to ensure that their surrogate decision resembles as closely as possible what the patient's *authentic* decision would be, i.e. what the patient would decide him- or herself. Consequently, it would seem that the better surrogates know the patient, the better are their chances of making a decision that is in line with how the patient would decide for him- or herself. Aside from the ensuing practical and epistemic problem of *how* to know the patient's self, the underlying theoretical challenge at this point is the well-known problem of spelling out the notion of a person's *self* in the first place, i.e. addressing the question of *who a person is*.

---

[1]For ethics as well as political philosophy in general, see the overview in Christman 2015, and for medical ethics in particular, see most prominently Beauchamp and Childress 2013, esp. ch. 4.

[2]With the only exception of cases of hard paternalism if one concedes that such acts can ever be ethically justified. See in this regard, for instance, Beauchamp and Childress 2013, 221–23.

[3]For an overview, see Jaworska 2017.

[4]Throughout my paper I use the term 'self' interchangeably with a person's 'identity' (as this term is usually understood in the social sciences), 'personality,' or 'character,' i.e. in general in terms of an answer to the everyday question *who* a person is and to spell out the criterion of *authenticity* within the notion of personal autonomy.

In this respect, *narrative accounts of the self* have gained popularity in recent decades.[5] The core idea is that we usually answer questions about who we are—if they are not merely aimed at finding out our name—by way of telling a story, and more precisely, our personal life story. Narrative accounts of the self then claim that these stories are, indeed, crucial to understanding the self. Knowing a person, therefore, basically means knowing her life story. Accordingly, knowing the patient's life story seems to be a solid foundation in order to make a surrogate decision that respects the patient's identity or self and also comes as close as possible to what the patient would decide for him- or herself, were he or she able to.

However, surrogates can obviously know a patient's life story only *thus far*, i.e. up to the present time in which the surrogate decision has to be made. The patient's life story is clearly not over yet, even if in cases of palliative care and end-of-life-decisions one might say that this is now its final chapter. Still, it seems to be an appealing idea for surrogates to use the patient's life story thus far and create the most fitting and coherent next or final chapter in order to respect the patient's autonomy. If, for example, a person's life story contains numerous occasions in which she struggled for independence and tried to solve any problem solely on her own, it seems only natural to say that the most fitting continuation of her story should retain this feature. Accordingly, the most fitting surrogate decision to honor this feature—which, ironically enough, will already unavoidably undermine it in this case—should at least aim at securing the greatest possible amount of future independence, for this would apparently be what the patient would cherish also in the situation at hand. Thus, narrative accounts of the self would have provided us with a simple and convincing solution to the problem of respecting patients' autonomy in surrogate decision-making.

However, in the following I will argue that this seemingly appealing solution is in fact neither simple nor convincing. To this end I will, firstly, introduce the idea of relying on narrative accounts of the self when it comes to surrogate decision-making in medical ethics in more detail. Secondly, I will distinguish between expressive narrative accounts, which merely rely on a patient's life story as an epistemic tool to find out about the patient's most probable authentic decision in the situation at hand, and constitutive narrative accounts, according to which a person's life story is, indeed, constitutive of the person's self. While expressive narrative accounts may indeed provide us with a helpful epistemic tool, they essentially presuppose a non-narrative account of the self as well as a contestable claim about stable character traits. Only constitutive narrative accounts will thus present an independent narrativist alternative to analyzing a person's self. However, I will, thirdly, argue that a person's narrative or life story is in a constant state of flux due to inherent characteristics of constitutive narrative accounts. Such accounts, therefore, not only fail in providing us with a reliable starting point from which to extrapolate *the most fitting* continuation

[5]Cp. MacIntyre 1981, ch. 15; Ricœur 1987, 1991, 1995, ch. 6; Schechtman 1996; DeGrazia 2005, ch. 3; Henning 2009; and Davenport 2012. For a concise overview, see Shoemaker 2016, Sects. 2.3 and 6.2. For seminal criticisms of narrative approaches to the self, see Strawson 2004; and Lamarque 2004.

of a patient's life story, but they also exclude the very idea of there being *a fitting continuation* in the first place. Hence, constitutive narrative accounts of the self are, unfortunately, also unable to present a simple and convincing solution to the problem of how to make a well-justified surrogate decision.[6]

5.2 Referring to a Patient's Life Story in Surrogate Decision-Making

Notably, Mark G. Kuczewski as well as Jeffrey Blustein have defended a narrative approach to surrogate decision-making.[7] Blustein "holds that proxy decision-makers should regard themselves as continuers of the life stories of those who have lost narrative capacity."[8] Kuczewski concurs with him: "The philosophical argument we advance is simple. People are their stories [...] This story has a prima facie moral claim to being finished in a way consistent with what has come before and with the values and self-conception with which the person has tried to shape it."[9]

The background theory of both Blustein and Kuczewski is essentially Marya Schechtman's narrative self-constitution view, "according to which a person creates his identity by forming an autobiographical narrative—a story of his life.[...] On this view a person's *identity* [...] is constituted by the content of her self-narrative, and the traits, actions, and experiences included in it are, by virtue of that inclusion, hers."[10] Yet, it should be noted that this only describes the core of Schechtman's view, as she immediately goes on to stress. A person's self-narration has to meet certain conditions in order to be plausible, namely an articulation constraint and a reality constraint.[11] The former is meant to avoid cases of self-deception or errors in one's self-view and to stress the need for a fitting narrative articulation of one's traits, although unarticulated attributes may still remain important features of the person's identity. The latter is meant to ensure that a person's self-narrative coheres with reality, i.e. it may not include factual errors or interpretive inaccuracies.

Assuming that a patient's self-narrative is plausible in these regards, it may then function as the crucial point of reference, based on which surrogate decision-makers can extrapolate a fitting continuation of the patient's life story and make their decision accordingly. This way, the surrogate decision would respect the patient's autonomy,

---

[6]To avoid a possible misunderstanding, I do not argue against either expressive or constitutive narrative accounts per se, i.e. in terms of such accounts being implausible analyses of the (constitution of) the self. I merely argue that neither of these accounts, convincing as they may be per se, is able to provide us with a solution to surrogate decision-making in medical ethics.

[7]See Kuczewski 1994, 1999; and Blustein 1999. See also Rich 1997; Lindemann Nelson 1997; McCarthy 2003; DeGrazia 2005, esp. ch. 3; Torke, Alexander, and Lantos 2008; and Witt 2018.

[8]Blustein 1999, 21.

[9]Kuczewski 1999, 33.

[10]Schechtman 1996, 93f. Cp. also DeGrazia 2005, 80–88, who roughly agrees with Schechtman's account.

[11]Cp. Schechtman 1996, 114–30; and Blustein 1999, 23 and 28.

since it would acknowledge the patient's narrative self-constitution and, consequently, acknowledge and respect who he or she is. Hence, turning to the patient's life story and extrapolating the most fitting continuation would apparently provide us with a simple and convincing answer to surrogate decision-making.

However, Blustein and Kuczewski readily admit that things are not that simple. Blustein concedes that his account "does not always yield unequivocal answers to questions about appropriate treatment"[12]—although he considers this a strength "because it is unreasonable to expect that proxy decision-makers will always be able to come up with uniquely correct answers in these situations."[13] Hence, his account is only meant to narrow the range of choices and exclude options that, more or less clearly, do not fit the patient's life story. Consequently, the treatment options available have to be put to the test of how well they fit the patient's life story.

> The test is not simple consistency, but how well a particular decision hangs together with elements of the formerly competent patient's identity-constituting narrative. Decisions that have a high degree of coherence with the rest of a patient's narrative self-conception are those that are faithful to who the patient was, as expressed in the organizing principles of her life through which she understood herself and her world. These decisions satisfy the condition of narrative fit and keep the patient's self-constituting narrative going after the loss of narrative capacity. […] Proxies do not have total creative freedom to interpret and continue the patient's life story in accordance with what they believe is in the patient's interest. […] Properly understood, the proxy's judgments are both free and constrained. They are free in the sense that they reflect, to some extent, the proxy's own authorial perspective; constrained in the sense that they can plausibly be regarded as a continuation of the patient's life story, not as the beginning of a new story.[14]

However, it should be noted that Blustein—and Kuczewski, too—tend to refer to "the" life story of the patient, apparently conceding that there is, indeed, only one or at the very least only one authoritative version of the patient's narrative self-constitution.[15] Hence, properly understood, Blustein and Kuczewski need to claim that there is, indeed, one authoritative life story *thus far* based on the patient's own narrative self-constitution. Yet, this story may have more than one sufficiently fitting continuation, which explains that there is, indeed, a range of possible options for the surrogate decision-maker to choose amongst, although the options are limited due to the criterion of sufficient narrative fitness. Still, the crucial claim of Blustein and Kuczewski is that a surrogate decision-maker should pick the treatment option that narratively fits and continues the patient's life story thus far *best*. Hence, although referring to the patient's life story may not be as simple as initially hoped, it would still provide us with a convincing account of how surrogates should make their decision in order to respect the patient's autonomy by way of respecting who he or she is.

---

[12] Blustein 1999, 22.

[13] Blustein 1999, 22, cp. also 24. Kuczewski agrees (cp. Kuczewski 1999, 33f.).

[14] Blustein 1999, 23. Kuczewski argues along the same lines (cp. Kuczewski 1999, 34).

[15] Cp. Blustein 1999, 25; and Kuczewski 1999, 33f. Given the possibility that the patient may have told multiple and more or less different versions of his or her life story, it remains an open question which of these versions should be seen as authoritative.

5.3 Narrative Accounts of the Self

However, so far, the narrative account of the self underlying the previous discussion remains somewhat unclear as to which role a person's values, preferences, or what she cares about play in comparison to the person's life story when it comes to the constitution of the person's self. While it is mostly claimed that a person's life story *constitutes* the person's self or identity, other passages by the respective authors seem to indicate that a person's life story could be merely an *expression* or articulation of the person's underlying self-constituting features, like her values, preferences, or what she cares about.[16] These two strands of narrative accounts and their implications should be distinguished more precisely. According to *expressive accounts*, a person's life story merely functions as a good or even the best way to express or describe a person's self, i.e. the underlying and self-constituting features. *Constitutive accounts*, on the other hand, hold that the self is, indeed, constituted by the story we tell about ourselves, i.e. they hold that the self *is nothing but the story* which is told about who we are.

The expressive account's central claim is, thus, rather weak and still compatible with different claims about the self's constitution.[17] In fact, expressive narrative accounts have to presuppose such a further account in order to determine the story's content in the first place. In contrast, constitutive narrative accounts are incompatible with such alternatives and present an independent approach to analyzing the self.

5.3.1 Expressive Narrative Accounts and Referring to a Patient's Life Story

Based on expressive narrative accounts, a patient's life story would, thus, not be a crucial or necessary element for how to decide which surrogate decision to make, for it would merely follow and describe the underlying constitutive features of the patient's self. It would, therefore, merely function as an epistemic tool to get to know and be able to acknowledge the patient's underlying values and preferences. If so, the patient's life story would in principle be replaceable by any other suitable epistemic means to gain insight into the patient's underlying constitutive features of his or her self and to determine what the patient's autonomous decision would probably be in the situation at hand.

---

[16]Although Schechtman's term for her view is obviously straightforward enough in claiming that a person's narrative is *constitutive* of her identity or self, the constraints she introduces seem to indicate that one's narrative may very well be conceived of as merely an adequate *expression* of an underlying content (cp. again Schechtman 1996, 114–30). Cp. also Kuczewski 1994, 43f., 1999, 33; Blustein 1999, 24–26; and DeGrazia 2005, 80.

[17]For a concise overview of general approaches to the constitution of the self, see Kühler and Jelinek 2013; and Shoemaker 2016.

Moreover, for the expressive narrative account to be reliable, it needs to presuppose that the patient's underlying constitutive features of his or her self, i.e. essentially his or her values, preferences, or what he or she cares about remain stable. Only then would it be possible to discern in a well-justified way which possible continuation would be the most fitting continuation of the patient's life story thus far. Otherwise, it would simply remain an open question if the surrogate decision-maker should refer to the patient's earlier values or preferences as expressed in his or her life story thus far.[18]

Taken together, expressive narrative accounts, therefore, not only fail to present an independent alternative to the constitution of the self, but they also have to presuppose the rather controversial claim that the person's relevant constitutive features of his or her self are stable character traits in order to be able to judge which possible continuation of the patient's life story would fit the story thus far best.[19] If so, the whole plausibility of the idea of referring to a patient's life story in order to make a suitable and well-justified surrogate decision hinges on this assumption of stability to begin with. Hence, expressive narrative accounts only succeed in providing a helpful epistemic tool for surrogate decision-making if we can safely assume that the patient's self-constituting features remain stable.

However, why should we take this assumption for granted? Given that we often enough come to reflect and reevaluate our preferences or what we care about in light of severe or tragic events, like in cases of serious medical predicaments, why should we assume that the patient's values and preferences remain unchanged and his or her life story will not take a substantial turn in light of such events? Assuming that such a possibility cannot be dismissed out of hand, surrogates lose the crucial criterion based on which they are intended to judge the fitness of any possible continuation of the patient's life story according to expressive narrative accounts. Consequently, expressive narrative accounts not only do not provide us with a simple answer to surrogate decision-making but also become substantially questionable in their justification of *fitting* continuations of patients' life stories.

5.3.2 *Constitutive Narrative Accounts and Referring to a Patient's Life Story*

As noted above, constitutive narrative accounts comprise the claim that a person's self is constituted by her life story,[20] which now also includes the idea that a patient's life story itself provides us with an answer to respecting the patient's autonomy and

[18]This problem essentially mirrors the similar problem with referring to a patient's advance directives, as one can never be sure that the patient's expressed preference in it may still be considered valid and has not changed since then.

[19]Blustein actually admits this presupposition of stability explicitly (cp. Blustein 1999, 22).

[20]Cp. MacIntyre 1981, 217f.; Ricœur 1995, 140f.; and Schechtman 1996, 94.

to who he or she is.[21] This way, constitutive narrative accounts are able to avoid the above criticism because the life story is no longer supposed to track any underlying other self-constituting and possibly changing features. Instead, it is the construction of the person's life story itself that determines who the person is and will be when it comes to the continuation of her story. Yet, this raises the question in what way the construction of a person's life story thus far may justifiably contribute to the decision on how it should continue, i.e. how it should be constructed when it comes to including the person's present situation and its bearing on her future.

Aside from Schechtman, Paul Ricœur is a prominent proponent of such a constitutive narrative account,[22] whose elaboration on it I find especially helpful for the purpose of this paper. Ricœur unfolds his narrative understanding of a person's self in three steps, with the first step being the identity of a story. In this regard, it has to be noted that single aspects of a story do not have meaning on their own but gain their meaning only in relation to each other and to the story as a whole. Accordingly, if we want to make sense of single aspects or situations in our life, we have to incorporate them in a story, i.e. we have to put them in a meaningful order and relate them to each other narratively.[23] Establishing such a meaningful order of events then constitutes the story's identity.[24]

In a second step, the story's identity is conveyed to its main character. Following Aristotle's *Poetics*, Ricœur argues that successful storytelling needs to achieve *concordance* or *coherence*. This means that those aspects of the main character's 'life' which are interpreted as *discordances* or *contingencies* regarding the kind of story told have to be incorporated plausibly as well. In order to do this, a fitting *configuration* of the story is needed, which is able to provide concordance or coherence on the whole. This configuration then also constitutes a concordant as well as dynamic identity of the story's main character. Taken together, the story and its main character display concordance or coherence and, thus, an intelligible narrative identity.[25]

In order to illustrate these two steps for my purpose at hand, imagine a woman leaving her house and throwing a teddy bear in the dumpster. First of all, this single

---

[21]Cp. Blustein 1999, 20f.; and Kuczewski 1999, 33.

[22]See Ricœur 1987, 1991, 1995, esp. ch. 6.

[23]On a critical note, Galen Strawson has prominently pointed out that putting aspects and situations of a person's life in a meaningful order does not necessarily mean telling a canonized form of story (cp. Strawson 2004, 439–43). Weaker versions of constitutive narrative accounts, thus, only comprise the claim that *some* form of meaningful order has to be established with regard to the aspects and situations in question. While this may be done in form of a canonized story, it does not have to be. In contrast, stronger versions of constitutive narrative accounts, like Ricœur's, claim just that. We need canonized forms of storytelling in order to construe a meaningful self. Moreover, we also need an adequate social context in which this may be done and from which the canonized forms of storytelling stem. Be that as it may, the crucial point for my discussion here is that both weaker and stronger versions of constitutive narrative accounts include the claim that at least some form of meaningful order needs to be construed in order to render single events or decisions intelligible within a coherent frame of reference.

[24]Cp. Ricœur 1987, 59–61, and 1995, 141–43.

[25]Cp. Ricœur 1987, 61–63, and 1995, 143–47.

action gains its meaning, and ultimately its meaning for who this woman is, only in relation to other actions, elements, or aspects of the woman's life. Assuming, for example, that the teddy bear was once a gift from her mother, and it represented the mother's controlling influence, we might see her throwing it in the dumpster as a way of gaining independence. Or, imagine that the woman originally gave the teddy bear to her own daughter, who, playing with it, died shortly afterwards in a car accident for which the woman wrongly blamed herself. Throwing the teddy bear away now might be interpreted as finally closing a tragic and guilt provoking chapter of her life and freeing herself from past burdens. In any case, any meaning or personal importance of throwing the teddy bear away can only be understood once this single action is put in relation to other elements of her life story and her life story as a whole, with the latter serving as the overall frame of reference and, thus, creation of meaning. Accordingly, the first version would be a story about emancipation, while the second version would come down to a story about moving on after personal tragedy.

Likewise, in order to understand the meaning of a medical situation a patient finds herself in, we have to take the same approach. Referring to the patient's life story thus far would allow us to understand the meaning of the current medical situation for the patient, and the importance of the decision which needs to be made, by way of interpreting the situation and decision as a fitting part of the patient's life story, thereby retaining overall coherency. Consider the following example: John has gotten in a freak accident and fallen into a coma. Although he is expected to wake up again soon, his right hand has been severely injured and needs medical attention immediately. He has previously given no indication of how he prefers to be treated in cases like this, which makes this a case for surrogate decision-making. The decision that now has to be made by his wife Jane, his surrogate decision-maker, is whether to try a rather new and risky procedure in order for John to regain full use of his right hand, but which could lead to a permanent numbness in his fingers if things do not go well, or to play it safe and opt for a tried-and-tested procedure which will ensure that John will retain feeling in his right hand and be able to use it again for everyday purposes, but which comes at the price of a significant loss of fine motor skills needed, for instance, for playing the piano.

In order to understand the situation's meaning for John, his wife would need to know whether and how taking risks as well as fine motor skills in his hands have featured so far in John's life story. Assuming that John's life story is neither one of a man who takes high risks for possible high rewards nor one of, for instance, a pianist (or including some other occupation or hobby for which fine motor skills are indispensable), it seems that Jane can safely conclude that losing fine motor skills in his right hand would not be connected in any especially meaningful or important way to John's major 'storyline.' Moreover, playing it safe would, indeed, be just one more expression of John's character arc in this respect, which began, let us assume, with some pretty risky behavior as a kid, which lead him to adopt an as risk-free lifestyle as possible later. Hence, it certainly looks like opting for the tried-and-tested procedure would be a fitting continuation of John's life story and, thus, the decision John himself would likely make.

However, Ricœur introduces one more and crucial step in the constitution of a person's self. This third and final step consists in the identification of the real person with the story's main character. This includes that we engage in a game of imagining certain variations of the story, including the main character's identity, and then identify with a specific version of the main character thus narratively defined. Furthermore, as we are always aware of this game of imagination and of the possibility of narrative variations, Ricœur claims that we may even gain new knowledge or a new awareness of ourselves. The varying stories may even be seen as a way of active self-interpretation and, thus, self-creation.[26]

If so, this means, firstly, that it necessarily remains an open question which meaning, if any at all, we attribute to a single action in relation to other narrative elements when we 'play' with different variations of the story. Secondly, this consequently also holds for the main character's identity, which is defined by each different variation anew. Hence, following Ricœur, a person's self may be constituted narratively, but only in terms of a specifically and actively construed variation of a concordant story with its main character for the person to identify herself with.

The general upshot is that there is not *one* authoritative life story, like Kuczewski and Blustein have to presuppose,[27] but many possible stories we create each time anew when 'playing' with different variations of *our* story and finally choosing one with which to identify. Accordingly, the idea of narrative self-constitution or being the author of one's own life story would imply that we likewise create our self anew each time we formulate a specific variation of our life story and its main character, thereby creating its specific identity, and then choose to identify with it. Moreover, the introduction of a choice in authoring our own life story and, thus, identity introduces even the idea of making an *autonomous* choice in this regard. In any case, following a constitutive narrative account like Ricœur's implies that there is substantial room for different life stories with different identities which substantially alter our narratively constituted self accordingly.

To illustrate this, we might imagine different versions of the woman's story mentioned above and relate the action of throwing the teddy bear in the dumpster to other actions or elements in her life. Maybe she forgot about the fact that it was once a gift from her mother, which is why she may not even include her throwing it away in her life story. And even if she remembered, we could easily imagine that she would not consider it symbolic of her relationship with her mother. The same holds with regard to John's situation. His wife Jane faces the same challenge in imagining and assessing different versions of John's life story. Could so far rarely mentioned instances of John attempting to play the piano or the fact that he took his fine motor skills always simply for granted now play a different and more meaningful or important role in some new version of his life story, especially given his current medical situation? It seems obvious that constructing respective versions of John's life story, which convey substantially different meanings to single events and decisions, resulting in a different identity of his overall 'storyline' or 'character arc' and,

[26]Cp. Ricœur 1987, 63–67, and 1995, 148–51.

[27]Cp. Blustein 1999, 25; and Kuczewski 1999, 33f.

thus, his self, are not only easily imaginable but may likewise easily be considered plausible according to constitutive narrative accounts.

In this regard, given that Ricœur emphasizes the importance of narratively constructing coherence by incorporating discordances into a coherent story, not even fundamental shifts or changes to John's life story thus far could be excluded when imagining how John's life story may look like now. For example, it is easily imaginable that John, were he awake, could fundamentally reevaluate his no-risk strategy in life in light of his current medical situation. Accordingly, the new version of his life story would be one in which he has grown into someone who has not only overcome his recklessness in his youth but also his far too restrained and risk-free lifestyle afterwards, now making him a person who can reasonably weigh risks against benefits and take careful decisions in light of what is at stake. How could Jane anticipate this based on his previous life story, if not only his story thus far but life stories in general always remain open to be constructed differently? On the contrary, according to constitutive narrative accounts, the challenge Jane faces as surrogate decision-maker does not even consist in anticipating the fitting continuation of John's life story and respecting it but in actively constructing John's narrative self by way of choosing one version of imaginable life stories in light of his past and his present situation. In this regard, constitutive narrative accounts necessarily miss any external criteria—barring basic descriptive coherence with reality—based on which any of these versions could be considered more or less justified. Essentially, Jane will, therefore, unavoidably become the surrogate author of John's narratively constituted self at this point.[28]

To sum up my line of thought, the following features of constitutive narrative accounts of the self, like Ricœur's, raise serious doubts that such accounts can provide us with a simple and convincing answer to the question of how to respect a patient's autonomy in situations of surrogate decision-making in medical ethics:

(1) We can choose with which variation of our story we want to identify.
(2) We are, thereby, the author of our own story.[29]
(3) We create our story each time anew when we tell a specific version of our story.
(4) In doing this, we not only create a specific story but also a specific meaning for each of its elements.
(5) We actively pick out certain elements and their relations to be included in our story, and, conversely, leave out others.

These aspects, i.e. essentially the ability to be, at least partially, the author of our own life story and, thus, of our own self, have crucial consequences for surrogate decision-making, for surrogates thereby unavoidably become surrogate authors of a patient's life story and, thus, of his or her self.

---

[28]While Blustein explicitly denies this (cp. Blustein 1999, 26f.), I take it that, given Ricœur's elaboration on constitutive narrative accounts and even Blustein's own partial admittance (cp. Blustein 1999, 27), his claim to the contrary remains unconvincing.

[29]This holds at least partially if the impact of social conditions on what may count as successful storytelling and which versions of a story a person comes to think of in the first place is taken into account.

While this may, indeed, be a valid implication of narrative accounts, it simply does not tell us anything about which decision the patient would autonomously make, i.e. with which version of his or her life story he or she might come up and identify with. Given the more or less infinite number of possible variations of the patient's life story, choosing one of them would surely not be a simple matter. Moreover, since no external justification on the basis of the patient's character is available to constitutive narrative accounts, choosing any one version over all others obviously remains unjustified. Hence, just like expressive narrative accounts, also constitutive narrative accounts are not only far from presenting a simple solution to surrogate decision-making but likewise remain unconvincing.

5.3.3 Narrative Self-Creation

Still, one might argue that the above criticism of constitutive narrative accounts hinges on a questionable understanding of authorship that includes a capacity for free or autonomous creation of the self. David DeGrazia has formulated this claim succinctly:

> Self-creation occurs when one doesn't just watch one's life story being written or look elsewhere as the chapters fly by; it occurs when an individual takes an active role in authoring the biography, making it a *lived* autobiography. … [T]he self-creator endeavors to write those later chapters – and perhaps, in light of the evolving story, edit earlier chapters as different themes emerge as critical and self-discoveries put old details in a different light. … It makes sense, then, to think of self-creation very roughly as *autonomous writing of self-narratives*.[30]

One might criticize that this claim crucially departs from constitutive narrative accounts of the self and instead comes down to an existentialist account of self-constitution, according to which we are radically free—or autonomous—in creating and determining our self, now in terms of authoring our own life story. In this regard, for instance Harry G. Frankfurt and Charles Taylor have argued that the existentialist position is untenable.[31] A radical choice that is not determined by anything at all also lacks any real connection to the person making it. Hence, it cannot plausibly be considered a choice of this person. Worse still, it even cannot be considered an intelligible choice at all. Consequently, the assumption of a radically free or autonomous self-creation is incoherent.[32]

Be that as it may, constitutive narrative accounts then face a dilemma when invoked for providing a simple and convincing solution to the problem of how to respect

[30] DeGrazia 2005, 107.

[31] Cf. Frankfurt 1993, 109f.; and Taylor 1977, 31–33.

[32] Of course, much more needs to be said about whether an existentialist approach to the constitution of the self is really untenable. For a concise overview of existentialism in general, see Crowell 2017, and for an existentialist approach to the constitution of the self in particular, see Kühler and Jelinek 2013, xif. and xviii–xx. For the purpose of my paper, however, I can leave this question open.

patients' autonomy best in cases of surrogate decision-making. Either narrative self-creation is considered impossible, then we cannot be the authors of our self, at least not in the sense of creating the content of one's life story. We are then merely—more or less truthful—conveyors of it as we discover who we are through the depiction of the story's main character. Consequently, our life story must be based on independently constituted features of our self, which have to be considered as given. Yet, if the contours of one's life story are considered given, i.e. the story's content, then this implies that one's life story simply has to track these underlying and given self-constitutive features.[33] If so, constitutive narrative accounts simply revert back to expressive narrative accounts, which, barring the questionable stability claim, have already proven to be unable to produce a convincing answer to how a patient's life story should continue.

Or narrative self-creation is considered possible, then a patient's life story thus far does not tell us anything about how it should continue. For, as noted above, a person's life story may also, and usually does, incorporate discordances or contingencies, including abrupt changes in the person's life and seemingly incoherent or arbitrary decisions. When incorporating these discordances in the story's overall configuration, a concordant or coherent identity is constituted, firstly of the story, secondly of the story's main character, and finally of the real person's self once the person identifies with the main character of this specific version of the story.

Accordingly, following Ricœur, freedom or autonomy in narrative self-creation consists in picking out the events and aspects of one's life which one may set in relation to each other as well as choosing a specific configuration of one's story, so that one determines the identity of one's life story and of oneself as its main character. Moreover, this freedom then also holds for others, and especially surrogate decision-makers, who may thus act as the surrogate author of the patient's life story at this point.

If so, practically any element or discordance can be integrated into a newly constructed configuration of the person's life story by way of relating them to other elements and thereby creating new overall coherence and a new meaningful frame of reference for these and all other narrative elements. Hence, while, for instance, a certain decision may appear arbitrary or completely out of character at first, its integration into the newly constructed person's life story may very well make it a coherent element of it.

Consider again the new version of John's life story mentioned above, which substantially departs from his previously constructed life story thus far. The decision to opt for the new and risky procedure may appear out of character at first, but only under the assumption that the identity of his life story remains stable. Yet, once again,

---

[33] Schechtman's discussion of the articulation constraint and reality constraint remains rather vague in this regard. If there are, indeed, any decisive criteria upon which one's self-narrative can be judged as being adequate or not, this would imply that one's life story has to track underlying and given content of one's life and character. However, if there are no such decisive criteria, this would imply that the constraints Schechtman advocates cannot plausibly go beyond simple descriptive coherence with reality, e.g. one's whereabouts at a certain time. Cp. Schechtman 1996, 115–19 and 125–30.

why should we take this assumption of stability for granted, especially in light of Ricœur's elaboration of constitutive narrative accounts? There is no good reason to do so, since the criterion of coherence with any previous version of John's life story cannot be invoked in constitutive narrative accounts because coherence is considered to be the result of narrative construction and can easily be achieved in new versions by way of suitably incorporating discordances, including discordances to previous versions of the story.

Consequently, if there is always the possibility in a person's life story of surprising new and discordant events, elements, or decisions, including seemingly arbitrary switches of preferences—all in comparison to previous versions of the story—there can obviously be no such thing as *narrative necessity* concerning how a person's life story has to be constructed now. Constitutive narrative accounts, thus, leave no room for the idea of *a fitting continuation* of a person's life story. Yet, this idea of a fitting continuation is needed in order to justify a specific surrogate decision as resembling as closely as possible the authentic and autonomous decision the patient would have made him- or herself based on an account of his or her self, which is why constitutive narrative accounts remain unconvincing also following the second horn of the dilemma.

5.4 Conclusion

I have argued that the seemingly appealing reference to a patient's life story when it comes to surrogate decision-making is in fact neither a simple nor a convincing solution. Expressive narrative accounts do not present an independent analysis of the constitution of the self to begin with and are at best a helpful epistemic tool to find out what the patient's autonomous preference likely would be. Yet, even this only holds under the contestable assumption that the patient's self remains stable. Constitutive narrative accounts, on the other hand, present an independent account of the constitution of the self. Yet, they either revert back to an expressive narrative account if the idea of one's free authorship of one's life story is rejected, or they fail to provide any reason for why one should choose a certain version of the patient's life story as the newly to be constructed one. Instead, a person's life story remains in a constant state of flux, which is why also constitutive narrative accounts are unable to provide surrogate decision-makers with a well-justified answer which treatment option to choose in order to respect the patient's autonomy based on who he or she is. In conclusion, referring to narrative accounts of the self when it comes to the problem of surrogate decision-making is, therefore, at the very least not the whole story and, as I have argued, not even a particularly crucial chapter.

References

Beauchamp, Tom L., and James F. Childress. 2013. *Principles of Biomedical Ethics*, 7th ed. Oxford: Oxford University Press.

Blustein, Jeffrey. 1999. Choosing for Others as Continuing a Life Story: The Problem of Personal Identity Revisited. *Journal of Law, Medicine and Ethics* 27 (1): 20–31.

Christman, John. 2015. Autonomy in Moral and Political Philosophy. In *The Stanford Encyclopedia of Philosophy*, ed. by Edward N. Zalta. Spring 2015. https://plato.stanford.edu/archives/spr2015/entries/autonomy-moral/.

Crowell, Steven. 2017. Existentialism. In *The Stanford Encyclopedia of Philosophy*, ed. by Edward N. Zalta, Winter 2017. https://plato.stanford.edu/archives/win2017/entries/existentialism/.

Davenport, John J. 2012. *Narrative Identity, Autonomy, and Mortality. From Frankfurt and MacIntyre to Kierkegaard*. New York: Routledge.

DeGrazia, David. 2005. *Human Identity and Bioethics*. Cambridge: Cambridge University Press.

Frankfurt, Harry G. 1993. On the Necessity of Ideals. In *Necessity, Volition, and Love*, by Harry G. Frankfurt, 108–16. Cambridge: Cambridge University Press, 1999.

Henning, Tim. 2009. *Person sein und Geschichten erzählen: Eine Studie über personale Autonomie und narrative Gründe*. Berlin: De Gruyter.

Jaworska, Agnieszka. 2017. Advance Directives and Substitute Decision Making. In *The Stanford Encyclopedia of Philosophy*, ed. by Edward N. Zalta, Summer 2017. Metaphysics Research Lab, Stanford University. https://plato.stanford.edu/archives/sum2017/entries/advance-directives/.

Kuczewski, Mark G. 1994. Whose Will Is It, Anyway? A Discussion of Advance Directives, Personal Identity, and Consensus in Medical Ethics. *Bioethics* 8 (1): 27–48.

Kuczewski, Mark G. 1999. Commentary: Narrative Views of Personal Identity and Substituted Judgment in Surrogate Decision-making. *Journal of Law, Medicine and Ethics* 27 (1): 32–36. https://doi.org/10.1111/j.1748-720X.1999.tb01433.x.

Kühler, Michael, and Nadja Jelinek. 2013. Introduction. In *Autonomy and the Self*, ed. by Michael Kühler and Nadja Jelinek, ix–xxxvi. Dordrecht: Springer.

Lamarque, Peter. 2004. On Not Expecting Too Much from Narrative. *Mind & Language* 19 (4): 393–408. https://doi.org/10.1111/j.0268-1064.2004.00265.x.

Lindemann Nelson, Hilde, ed. 1997. *Stories and Their Limits: Narrative Approaches to Bioethics*. Routledge.

MacIntyre, Alasdair. 1981. *After Virtue: A Study in Moral Theory*. Notre Dame: University of Notre Dame Press.

McCarthy, Joan. 2003. Principlism or Narrative Ethics: Must We Choose Between Them? *Medical Humanities* 29 (2): 65–71.

Rich, Ben A. 1997. Prospective Autonomy and Critical Interests: A Narrative Defense of the Moral Authority of Advance Directives. *Cambridge Quarterly of Healthcare Ethics* 6 (2): 138–147. https://doi.org/10.1017/s0963180100007763.

Ricœur, Paul. 1987. Narrative Identität. *Heidelberger Jahrbücher* 3: 57–67.

Ricœur, Paul. 1991. Narrative Identity. *Philosophy Today* 35 (1): 73–81.

Ricœur, Paul. 1995. *Oneself as Another*. Trans. by Kathleen Blamey. Chicago: University of Chicago Press.

Schechtman, Marya. 1996. *The Constitution of Selves*. Ithaca: Cornell University Press.

Shoemaker, David. 2016. Personal Identity and Ethics. In *The Stanford Encyclopedia of Philosophy*, ed. by Edward N. Zalta, Winter 2016. Metaphysics Research Lab, Stanford University. https://plato.stanford.edu/archives/win2016/entries/identity-ethics/.

Strawson, Galen. 2004. Against Narrativity. *Ratio (New Series)* 17: 428–452.

Taylor, Charles. 1977. What Is Human Agency? In *Philosophical Papers 1. Human Agency and Language*, by Charles Taylor, 15–44. Cambridge: Cambridge University Press, 1985.

Torke, Alexia M., G. Caleb Alexander, and John Lantos. 2008. Substituted Judgment: The Limitations of Autonomy in Surrogate Decision Making. *Journal of General Internal Medicine* 23 (9): 1514–1517. https://doi.org/10.1007/s11606-008-0688-8.

Witt, Karsten. 2018. 'Narrative and Characterization'. *Erkenntnis*, 1–19. https://doi.org/10.1007/s10670-018-0017-5.

Michael Kühler is Assistant Professor at the University of Twente, Netherlands, and "Privatdozent" (roughly equaling Associate Professor) at Münster University, Germany. His research interests include ethics, medical ethics, metaethics, and political philosophy. His recent publications include "Technological Moral Luck," in: Beck, Birgit/Kühler, Michael (eds.): *Technology, Anthropology, and Dimensions of Responsibility,* Techno:Phil 1, Stuttgart: Metzler, 2020, 115–132, URL: https://doi.org/10.1007/978-3-476-04896-7_9https://doi.org/10.1007/978-3-476-04896-7_9, "Romantische Liebe und die Freiheit zu gehen," in: Dietz, Simone/Foth, Hannes/Wiertz, Svenja (eds.): *Die Freiheit zu gehen. Ausstiegsoptionen in politischen, sozialen und existenziellen Kontexten*, Wiesbaden: Springer, 2019, 195–222, "*Modus Vivendi* and Toleration," in: Horton, John/Westphal, Manon/Willems, Ulrich (eds.): *The Political Theory of Modus Vivendi*, Dordrecht: Springer, 2019, 235–253, "Toleranz und/oder Paternalismus im engeren sozialen Nahbereich?," in: *Zeitschrift für Praktische Philosophie* 4, 2017, 63–86; and "What if I Cannot Do What I Have to Do? Notions of Personal Practical Necessity and the Principle 'Ought Im-plies Can'," in: Bauer, Katharina/Mieth, Corinna/Varga, Somogy (eds.): *Dimensions of Practical Necessity. "Here I Stand. I Can Do No Other,"* Houndmills: Palgrave Macmillan, 2017, 87–107.

Chapter 6
The ART of Authenticity

Birgit Beck

Abstract Recent progress in biomedical research in the fields of stem cell technology and genome editing has entailed a revival of ethical concerns regarding the autonomy and authenticity of future persons who might be created by prospective novel means of assisted reproductive technologies (ART), e.g., on the basis of in vitro gametogenesis or germline interventions. In this regard, critical authors refer to the Habermasean concern that persons "made" by technological means instead of coming into existence in a "natural" way could be deprived of forming a self-concept of being autonomous and the authentic "authors of their life histories". Since the recent birth of two genetically modified girls, these concerns have gained topicality. The notion of being the "author" of one's life history appears to relate to the idea that we can—somehow and within certain limits—actively influence who we are. It is hard to see why this idea could not include the possibility of integrating the circumstances of one's coming into existence—be they "artificial" or "natural"—in an authenticity-preserving way. Given that (future) persons are (will be) aware of these circumstances, they should be able to adapt their life histories accordingly. Assuming a relational structure of autonomy and authenticity in general, it appears plausible that no person's life history and self-concept are constituted and maintained in an isolated and individualistic manner. Quite the contrary, (self-)ascriptions of autonomy and authenticity require adequate social relationships, in the first place. Based on this assumption, it can be supposed that deprivation of a confident self-concept and feeling of "objectification" by their "creators" in (future) persons conceived by means of novel ART would be due to flawed communicative relationships or prejudice rather than an actual lack of essentialist authenticity. The Habermasean concern can thus be regarded as revealing a structural social problem rather than one raised by biotechnology per se, and it can be smoothed out by Habermas' own conceptual conditions for the development of personal identity, authenticity and autonomy.

B. Beck (✉)
TU Berlin, Berlin, Germany
e-mail: birgit.beck@tu-berlin.de

M. Kühler and V. L. Mitrović (eds.), *Theories of the Self and Autonomy in Medical Ethics*, The International Library of Bioethics 83,
https://doi.org/10.1007/978-3-030-56703-3_6

6.1 Pandora's Box: From Bioethical Thought Experiments to Real-World Problems

In November 2018, one day before the start of the *Second International Summit On Human Genome Editing* in Hong Kong, the scientific community was appalled by the announcement of biophysicist He Jiankui from the Southern University of Science and Technology of China in Shenzhen, who revealed via video message and in an Associated Press interview the successful manipulation of human embryos with CRISPR-Cas9 in the course of assistant reproductive treatment and the birth of apparently healthy twin-girls, whom he declared to be (in one case partly) resistant to HIV.[1] He reported that he had manipulated the CCR5 genes of 11 embryos stemming from 7 couples, the men of whom were HIV carriers and who had consented to take part in an experimental "'AIDS vaccine development' program".[2] Subsequently, 6 implantations were conducted, one of which resulted in a successful pregnancy.[3] The twins, whom He referred to as "Nana" and "Lulu", have the dubious honour of being the world's first "designer babies".[4] Although He declared in the said interview that he "felt a strong responsibility" and his aim was "to help the families and their children", and although he tried to justify his conduct at the summit, the message unsurprisingly caused worldwide disturbance and harsh critique. The university distanced itself from the "deeply shocking" experiments and 122 Chinese scientists published an open letter in which they protested against "opening Pandora's box", expressing their fear that this incident could cause serious harm for the reputation of Chinese science. Apart from "widespread condemnation by the summit organizing committee, world scientific academies and prominent scientific leaders",[5] several spokespersons of (inter-)national ethics committees brought forward their concerns, for example, the (former) chairman of the German Ethics Council, Peter Dabrock, who stated that the creation of manipulated children "would be a worst case scenario" and that He's announcement one day before the summit could "only be considered to be an affront against the aim of responsible science".[6] The chairwoman of the European Group on

---

[1]Cf. Regalado (2018) and https://apnews.com/4997bb7aa36c45449b488e19ac83e86d (accessed: 30 July 2020).

[2]Ibid. It has been questioned if the concerned prospective parents were really aware of what they were consenting to. Moreover, it has been criticised that manipulating the CCR5 gene with the aim of causing HIV resistance may have serious side effects and cause interdependencies like increased susceptibility to other infectious diseases. Generally, the danger of genetically manipulated innate immunity leading to "collective action problems" has been critically assessed previously; cf. Gyngell and Douglas (2015, 244 f.).

[3]In January 2019, Chinese authorities confirmed a second pregnancy resulting from the experiments.

[4]Admittedly, the notion of "designer babies" is not clearly defined; cf. Segers et al. (2019, 500): "There is relative consensus about the idea that the creation of designer babies involves genetic interventions in order to influence the traits of future offspring, but this is vague, and it can be rightly asked which interventions and which traits one is talking about."

[5]Dzau et al. (2018, 1215).

[6]https://www.br.de/nachrichten/wissen/erste-genveraenderte-babys-china-genome-editing-crispr-cas,RAUwtIy (accessed: 26 July 2019; own translation).

Ethics in Science and New Technologies, Christiane Woopen, took the same line and accused He and his team of having "offended human rights and inflicted serious harm on the trustworthiness of science. The international community should not tolerate this."[7]

Of course, the scientific community was neither inclined to tolerate this nor prepared for doing so (aside from decades of dealing with pertinent bioethical thought experiments theoretically).[8] However, the community had apparently been all too self-secure that none of their members would be courageous—or inconsiderate—enough to *seriously* attempt to break the seemingly unequivocal and generally accepted taboo of intervening in the human germline at this early stage of scientific development—for plain reasons of high insecurity regarding unforeseeable side effects, untenable risks for future generations (and, very probably, scientific suicide instead of fame).[9] As early as in 2015, in the forefront of the first *International Summit On Human Gene Editing* in Washington, D.C., renowned scientists in the field, among them Jennifer Doudna, one of the co-discoverers of CRISPR-Cas,[10] had called for "[a] prudent path forward for genomic engineering and germline gene modification"[11] and even for a moratorium on research in human germline modification, following the lead of the famous Asilomar conference in 1975.[12] However, at the first summit the call went unheard, only to be taken up again recently[13]—a little too late.

The whole debate on human genome editing and possible germline interventions gained momentum when, in 2015 for the first time, a group of scientists conducted experiments on non-viable human embryos for basic research of possible future therapies for the blood disease β-thalassaemia.[14] Although these studies did not include the remote possibility of creating live offspring and thereby manipulating future

[7]Ibid. (accessed: 26 July 2019; own translation).

[8]The community *could* in fact have been prepared, had some well-informed persons, e.g., He's former doctoral adviser, at times blown the whistle on his experiments; cf. Lander et al. (2019, 166): "[S]cientists who were apparently aware of this work did not take adequate measures to stop it." Interestingly, since He's fall from grace, even staunch liberal and transhumanist advocates of genetic enhancement (e.g. Savulescu and Kahane 2009; Savulescu et al. 2015) climb down from arguing too optimistically in favour of germline interventions, all the more since it has become apparent that not only overenthusiastic scientific renegades like He might have a share in "playing God", but also so far deemed rather innocuous DIY-biohackers; cf. Regalado (2019).

[9]Cf. Ledford (2015, 311): "No one so far has declared an interest in producing live babies with edited genomes, and initial experiments would suggest that it is not yet safe. But some suspect that it is only a matter of time"; Hildt (2016, 2): "[C]urrently, in view of the high risks and unresolved practical, societal and ethical issues involved, nobody would ever argue toward attempting human germline modifications that involve initiating a pregnancy. However, the move toward in vitro human germline interventions attempts to set the stage for future uses involving human reproduction."

[10]Cf. Doudna and Charpentier (2014).

[11]Baltimore et al. (2015).

[12]Cf. Lanphier et al. (2015).

[13]Cf. Lander et al. (2019).

[14]Cf. Liang et al. (2015), Cyranoski and Reardon (2015).

generations and have been assumed to be ethically unproblematic in this regard,[15] in reaction to them, well-known slippery slope arguments concerning "eugenics"[16] and the creation of "designer babies"[17] were revived in the subsequent bioethical and public discussion "against the background of a long interdisciplinary debate on ethical and societal issues of germline interventions that began in the 1980s".[18]

> In this debate, a broad spectrum of topics has been discussed, including medical issues concerning safety and efficacy, possible health benefits, risks and unintended consequences for future generations, and ethical issues such as individual and collective responsibility, reproductive autonomy, human dignity, the moral status of human embryos, human nature, tampering with human evolution, playing God, eugenics, discrimination, human enhancement, and transhumanist visions. [...] Up to a few months ago, in the absence of a technique to modify the human germline by disabling, inserting, or replacing DNA fragments, the debate on human germline interventions seemed to be rather academic and theoretical, far away from any reality. It is the advent of this new technique of gene editing and the observation that it can in principle be used to modify the DNA of the resulting organism and its offspring that revitalizes the debate on germline interventions and gives it a new direction.[19]

Invoking scenarios of human "hubris", "playing God", unduly interfering with "human nature" and shopping in the "genetic supermarket"[20] has frequently been considered to be a crypto-religious or science fiction, in any case, inconclusive line of argument by advocates of bioliberal positions,[21] and the rebuttal of such arguments as flogging a dead horse.[22] Unfortunately, those arguments have turned out to be not as far-fetched as they seemed, as the above-mentioned case of He, "Lulu" and "Nana" has demonstrated. More sophisticated versions of arguments against germline modification and genetic enhancement frequently refer to Jürgen Habermas' essay *The*

---

[15]Cf., e.g., Savulescu et al. (2015, 477).

[16]Cf. Pollack (2015), Kipke et al. (2017).

[17]Cf., e.g., Ledford (2015), Lander (2015, 5), Savulescu et al. (2015, 477), Baker (2016, 272 f.), Baumann (2016, 141); for a discussion of the possible creation of designer babies in the context of in vitro gametogenesis, see Segers et al. (2019).

[18]Hildt (2016, 1 f.).

[19]Ibid., 2; cf. also Kipke et al. (2017, 250): "The problem [with genome editing] lies in what becomes possible and probable with it: arbitrary interventions in the human germline, the irreversible transfer of alterations to future generations, serious shifts in the relation between generations, ethically dubious clinical experiments and the blurring of species borders. The rapid, even precipitous implementation of gene editing in Great Britain [the authors refer to the legalisation of mitochondrial transfer] for – up to now – experimental aims broaches at a single blow all the concerns that have been formulated regarding germline manipulation, eugenics and designer babies in the past decades" (own translation).

[20]Cf. Nozick (1974, 315), Gyngell and Douglas (2015).

[21]For example, a lot of criticism concerning arguments about hubris, playing God and human nature has been passed on the President's Council on Bioethics (2003) report *Beyond Therapy*.

[22]Thomas Gutmann, philosopher of law, member and spokesman of the former Centre for Advanced Study in Bioethics at the University of Münster in Germany, once kidded in a public talk that there are "zombie arguments" in the bioethical debate—bad arguments which reappear regularly and cannot be "killed" because they are already "dead". Until recently, the argument about designer babies appeared to be such a zombie argument.

Future of Human Nature,[23] in which he assumes that, apart from obvious problems of medical risks and side effects, the deeper problem of manipulating future children's genome is that this makes them very likely to suffer the loss of personal identity, authenticity and autonomy.

6.2 The Habermasean Argument

Since the first publication of Habermas' essay, his argument against genetic enhancement has been interpreted in a variety of ways and has drawn considerable criticism from different scholars. Jonathan Pugh, one of Habermas' more complaisant critics, states that "[r]esponses to Habermas in the bioethical literature have been, in the main, dismissive, concise and even hostile."[24] He assumes that Habermas' point has frequently been misunderstood and that "liberal bioethicists have thus far failed to provide a decisive objection to Habermas' arguments against germline enhancements."[25] In order to find a suitable starting point of critique, Pugh reconstructs Habermas' relevant notion of autonomy. According to Pugh, Habermas assumes that individual personality, authenticity and autonomy are influenced by both the natural conditions of being born and embodied and social conditions of communication and reciprocal recognition.[26] Despite its dependency on natural as well as social conditions, individual personality is wholly determined neither by internal nor by external forces "as long as the agent has a conception of their own self that they can identify with in abstraction from these determining influences."[27] This conceptualisation obviously requires the notion of a substantial "self", which in turn presupposes an essentialist approach of authenticity.[28] Although the supposed self "existed prior to the forces of socialization",[29] which, according to Pugh, is the reason for the person's ability to detach herself from her "socialization fate",[30] this does not mean that Habermas falls prey to the "fallacy of genetic determinism"[31] since it is crucial for his argument that the person be able to distance herself from her "natural fate"

[23]Habermas (2003). The original German version, *Die Zukunft der menschlichen Natur. Auf dem Weg zu einer liberalen Eugenik?*, was first published in 2001 and reprinted in 2002; an extended version was published in 2005. Interestingly, Habermas states explicitly that he is *not* concerned about germline interventions in his essay (cf. Habermas 2003, 117, endnote 2). Nonetheless, the Habermasean argument has attracted attention again in the recent debate on genome editing and germline interventions.

[24]Pugh (2015, 145).

[25]Ibid.

[26]Cf. ibid., 146 f.

[27]Ibid., 146.

[28]For an extensive discussion of different approaches of authenticity, see Guignon (2004).

[29]Pugh (2015, 147).

[30]Ibid., 146.

[31]Ibid., 149.

as well.[32] In reference to Arendt's concept of "natality", the moment of birth, for Habermas, marks "a new beginning" and "the onset [...] of a differentiation between the socialisation fate of a person and the natural fate of her organism".[33] Habermas takes the tension between biological contingency, i.e. the fact that a person's natural origin has been beyond our disposal so far, and social constitution to be a necessary condition for personal autonomy.[34]

Pugh depicts two lines of argument from Habermas' essay, terming them "the empirical argument" and "the argument from natality".[35] The empirical argument is concerned with a genetically modified child's lacking capacity to detach herself from her "socialization fate". While parents usually impose their visions of a good life and corresponding values on their child via education, the case of genetic intervention differs from this common practice since the genetically modified child literally "*incarnate[s]*"[36] the respective values and expectations, which impedes her ability to distance herself from them by way of reflective and communicative engagement. The argument from natality, in turn, states that depriving a child of a contingent biological origin makes it impossible for her "to conceive of herself as an autonomous agent".[37]

In a recent article on "Reproductive Medicine and Parental Responsibility",[38] Tatjana Tömmel restates Habermas' argument in the following way:

> Habermas's starting point is the observation that genetic interventions already possible today have blurred the once clear distinction between what has 'grown' and what is 'made' (Habermas 2005, p. 45). The former unavailability and unpredictability of a child's genetic features is now potentially placed at their parent's (or another adult's) disposal (Habermas 2005, p. 29). Habermas claims that the original biological contingency might be the necessary condition for being oneself, and for the egalitarian nature of our interpersonal relationships (Habermas 2005, p. 29). On this basis, he argues that the knowledge about the manipulation of one's own genetic features might undermine both the way the individual can lead her life autonomously, and of the symmetric relationships between free and equal persons (Habermas 2005, p. 45).[39]

The core problem with manipulating a child's genetic make-up, according to Tömmel's interpretation of the Habermasean argument, lies in a flawed parent-child relationship, which usually "should be characterized by unconditional acceptance, transparency, and the possibility of (future) equality and recognition".[40] Even granted

---

[32] Ibid., 146.

[33] Habermas (2003, 59).

[34] Cf. ibid.: "It is only by referring to this difference between nature and culture, between beginnings not at our disposal, and the plasticity of historical practices that the acting subject may proceed to the self-ascriptions without which he could not perceive himself as the initiator of his actions and aspirations."

[35] Pugh (2015, 147).

[36] Ibid. (original emphasis).

[37] Ibid., 147.

[38] Tömmel (2020) (the references in the quotes refer to the extended German version of Habermas' essay).

[39] Ibid., 181.

[40] Ibid., 183.

that prospective parents sincerely care about the well-being of their future children and make use of genetic interventions with the aim of providing them with the "best chance of the best life",[41] there is a danger that parents "create a new communicative situation. Instead of interacting with their (future) child in the role of a second person, they treat them as an object or 'malleable product' designed to fit their own wishes (Habermas 2005, p. 30)."[42] It is this supposed objectification of children that threatens their self-concept as autonomous persons and the development of adequate social relationships:

> According to Habermas, this attitude takes away from children the right to be *authors* of their own lives. Thereby it would undermine the possibility of a future relationship of mutual recognition, in which the other person is accepted in her contingent suchness. Habermas concludes that with the dissolution of the clear demarcation between persons and objects, a previously unknown interpersonal relationship would develop. Such a relationship would form an alien element within the principally egalitarian relationships, which are based on mutual recognition and which are legally instituted in modern societies (Habermas 2005, p. 30).[43]

To provide a third and last interpretation, in a statement against genome editing research on human embryos published before the actual realisation of germline editing, Roland Kipke, Markus Rothaar and Martin Hähnel, presciently arguing that *prima facie* ethically unobjectionable basic research could pave the way for premature clinical implementation or even eugenic practices,[44] refer to Habermas' argument:

> Manipulating the offspring's genetic make-up at will is likely to lastingly shape the relation between parents and children in a negative way. The child becomes a *made* instead of a *grown* being. They would face their parents not only as creators of their existence, but as planners of their physical and mental endowments. Parents would have to justify their eugenic decisions to their children, the children's life would always be subjected to the goal of the fulfilment of the parental plan. Compared to classical measures of influence by education, this targeted technological manipulation would have a new dimension. Facing each other at eye level, a relation of free and equal persons, which is desirable between adult children and their parents, is likely to be considerably impeded.[45]

Such an outlook appears indeed seriously worrying, now that genetically modified children no longer belong to the realm of bioethical science fiction thought experiments. If one supposes that the Habermasean argument is conclusive, it might turn out that the above-mentioned twins "Nana" and "Lulu" will not be able to (i) conceive of themselves as autonomous persons and authentic authors of their life histories and to (ii) engage in a free and equal relationship with their parents—or any other not genetically manipulated persons—once they are mature enough to understand and

[41] Savulescu and Kahane (2009), cf. also Savulescu (2001).

[42] Tömmel (2020, 181).

[43] Ibid., 182 (original emphasis).

[44] Cf. Kipke et al. (2017, 251 f.): "The assumption that one will stick to this kind of [fundamental] research is naive" (own translation).

[45] Ibid., 250 (original emphasis; own translation).

reflect on the circumstances of their coming into existence. Therefore, for the sake of these girls and other children like them who are likely to be born in the not so remote future,[46] it might be crucial to find a way to show that the Habermasean concern is implausible.

6.3 The Habermasean Argument in the Case of "Nana" and "Lulu"

One way to reject the Habermasean concern regarding personal authenticity and autonomy as well as egalitarian relationships in the case of "Nana" and "Lulu" could be to deny that they are in fact "designer babies" by adopting the following definition of Segers et al., who specify

> the creation of designer offspring as genetic interventions by means of editing and/or selection techniques in order to create offspring with nondisease related traits which match the future parents' preferences and which might, but need not be, eugenic.[47]

Assuming that their parents have consented to the intervention for "disease-related" reasons and not with the aim of realising arbitrary preferences, one could argue that the girls do not fall under the category of "designer babies", in the first place. Leaving this terminological issue aside, one could still insist that both concerns, the lack of being able to conceive of oneself as an autonomous person and the authentic author of one's life history as well as to engage in reciprocal egalitarian relationships, only hold in cases of germline manipulations for enhancement purposes, and not in cases of health-related interventions.[48] In this regard, Pugh points out that, in reference to the standard distinction between therapy and enhancement,[49] Habermas does not explicitly reject genetic interventions in principle, as long as one can reasonably, albeit counterfactually, assume that the future child would give consent to the measure: "He claims that therapeutic genetic modifications are permissible because we may presume consent on behalf of a modified individual to avoid a profound evil which is 'unquestionably extreme, and likely to be rejected by all'".[50]

---

[46]Recently, the Russian scientist Denis Rebrikov has announced to prepare genome editing experiments with the aim of preventing inherited deafness.

[47]Segers et al. (2019, 501).

[48]Tömmel (2020, 181 f.) emphasises the same point: "While negative eugenics prevent 'highly generalized evils' such as fatal diseases, positive eugenics select or 'enhance' certain traits according to the subjective preferences of the embryo's parents[.] [...] Thus, not reproductive medicine in general but selecting or manipulating the embryonic genome without any serious indication would be a clear distortion of power between parents and children."

[49]Of course, this distinction has been frequently criticised for relying on vague and contested concepts of health, disease and normality, respectively; Habermas (2003, 51) defends the contested differentiation: "However hard it may be to distinguish in the individual case between therapeutic interventions—the prevention of evils—and enhancing interventions, the regulative idea that governs the intended delimitation is simple."

[50]Pugh (2015, 148).

However, in the case of the above-mentioned intervention, it is doubtful if lacking innate resistance to HIV can be considered to be an "unquestionably extreme evil which is likely to be rejected by all". Contracting HIV is certainly an evil which is commonly rejected. Susceptibility to contracting HIV is, however, a disposition which is still naturally shared by many persons (nonetheless, this disposition can certainly be regarded as evil). Moreover, it has been questioned if today, given that there are adequate preventive measures as well as effective medication, this evil is so "extreme" that it can justify a high-risk and medically dubious germline intervention.[51] In fact, some critics have suspected that He's experiments did not have any therapeutic intention in the first place, which appears plausible since the future children were not expected to be carriers of the disease and the intervention was explained to the prospective parents as a "vaccination" programme. Vaccination (if genome editing can be compared to this measure at all), however, is usually considered to fall into a grey area between therapy and enhancement. Therefore, it appears unconvincing to justify this instance of germline genome editing by referring to the—in any case controversial—dichotomy between therapy and enhancement. Moreover, even consideration of the manipulation as a therapeutic intervention would not help to smooth out the Habermasean concern. After all, Pugh states that

> [i]f the contingency of one's genome is necessary for one's sense of self continuity, and *a fortiori* autonomy, then this seems to imply that a child who is the subject of a therapeutic genetic intervention, and whose genome is thereby not contingent, cannot have the sense of self-continuity that is necessary for autonomy.[52]

So far, the strategy of rejecting the Habermasean argument has not been successful, irrespective of the conceptualisation of the above-mentioned case as an instance of therapy or enhancement. The key to resolving the problem rather appears to lie in the notion of (essentialist) authenticity and autonomy, respectively.

6.4 Being the Author of One's Life History

Spelling out Habermas' argument from natality, Pugh lays emphasis on the notion of a substantial "self" which has to exist independent from social influences. He takes this to mean that

> in order for an agent to regard himself as the master of the social forces that presently determine his self-constitution, he must have a conception of himself existing in abstraction from the determination of these forces *before* he was born, and consequently subjected to these forces.[53]

[51] Cf. Regalado (2018): "Still, removing the CCR5 gene to create HIV resistance may not present a particularly strong reason to alter a baby's heredity. There are easier, less expensive ways to prevent HIV infection. Also, editing embryos during an IVF procedure would be costly, high-tech, and likely to remain inaccessible in many poor regions of the world where HIV is rampant."

[52] Pugh (2015, 149) (original emphasis).

[53] Pugh (2015, 148) (own emphasis); cf. also ibid., 146 f.: "In appealing to the concept of natality, Habermas is able [to] account for the possibility of autonomy despite the forces of socialization,

Genetic manipulation—as explained above, irrespective of therapeutic or enhancing aims—would "threaten the child's sense of a self-continuity between himself as he exists as a product of socialization, and himself as he existed in abstraction from this process, *prenatally*."[54] However, this cannot be the crucial point of Habermas' argument. The possibility of referring retrospectively to "me as an embryo" (before birth) is a necessary condition for *both* conceiving of oneself as a person with an unaltered genome *and* as a person whose genome has been manipulated. Habermas' concern is not a problem of *persistence*, in the sense of diachronic personal identity, although he observes that "the person can only see himself as the author of ascribable actions and as the source of authentic aspirations if he assumes continuity for a self, remaining self-identical in the course of a life history."[55] However, diachronic identity (as *one and the same organism*, irrespective of its genetic make-up)[56] must apply to both genetically manipulated and non-manipulated persons. Accordingly, the meaning of the concept of "personal identity" that Habermas puts forth must differ from biological persistence. The "self" he refers to is an *abstract* point of reference "beyond the lines of tradition and the contexts of interaction which constitute the process of formation through which personal identity is moulded in the course of a life history."[57] Here, the meaning of the term "beyond" cannot be *prior* to socialisation, as Pugh assumes. The authentic "self" is not an essential spatio-temporal entity in a literal sense. It is rather a *regulative idea*, a *transcendental subject* in the Kantian sense, so to speak.[58] Conceptualisation of the problem as one of persistence amounts to a category mistake. In contrast, what Habermas denotes as "personal identity" is better conceptualised as *personhood* and *personality*. The properties of personhood and personality are, however, "essentially constituted through social relations, which take place in recognition relations."[59] Habermas explicitly endorses this conceptualisation:

> It takes entrance in the public sphere of a linguistic community for a natural creature to develop into both an individual and a person endowed with reason. [...] In the symbolical network constituted by the relations of mutual recognition of communicatively acting persons, the neonate is identified as 'one of us.' He gradually learns to identify himself

---

since he claims that our natality allows us to identify the 'self' as it existed prior to the forces of socialization."

[54] Ibid. (own emphasis).

[55] Habermas (2003, 59).

[56] Cf. Quante (2018, Chap. 2). Quante plausibly differentiates four meanings of the concept "personal identity" that should not be confounded: *synchronic identity* (being one person at a given point in time), *diachronic identity* or *persistence* (numerical identity over time), *personhood* (the moral status of being a person), and *personality* (the individual characteristic of personhood).

[57] Habermas (2003, 59).

[58] The reference to Kant may, however, be misleading insofar as, for Habermas, even subjectivity is a product of social interrelations: "Subjectivity, being what makes the human body a soul-possessing receptacle of the spirit, is itself constituted through intersubjective relations to others" (Habermas 2003, 34).

[59] Quante (2018, 29 f.).

– simultaneously as a person in general, as a part or a member of his social community (or communities), and as an individual who is unmistakably unique and morally nonexchangeable."[60]

If it is true that personhood is constituted by (self-)ascription through social relations of mutual recognition and that personality is acquired and constantly "updated" in the process of engaging in communicative relationships, in the first place, then it appears that developing an authentic sense of self is also something that we *learn* by taking part in social relationships. Forming the regulative idea of a substantial and diachronically self-identical "self" which underlies one's own personality and functions as a point of reference for ascriptions of personhood and autonomy is itself dependent on interpersonal communicative relations in which an according socialisation takes place. From being endowed with an "all-natural" genome alone, no one comes up with the idea of possessing, let alone being an essential "authentic self". Accordingly, from being endowed with a manipulated genome alone, no one necessarily comes up with the idea of being objectified or lacking an essential "authentic self", *as long as they are not (if only implicitly, by a tacit social normative framing) taught to do so*. If, however, genetically manipulated children grow up aware of the fact that their fellow humans hold the belief that "the conditions [...] of nature-like growth [...] alone allow us to conceive of ourselves as the authors of our own lives and as equal members of the moral community",[61] *it is no wonder* if they internalise the detrimental belief that they are deprived of full-fledged personhood and unable to conceive of themselves as authentic authors of their life histories and autonomous persons.[62] In this regard, I agree with Tömmel, who notes that Habermas (ironically) "seems to underestimate the power of socialization".[63] Just like "no man is an island", no person can be considered to be the *sole* author of her life history. No person's self-understanding is constituted and maintained in an isolated and individualistic manner. As Charles Guignon notes, authenticity is a "*social* virtue".[64] Development of a consistent self-concept requires, first and foremost, adequate social relationships of mutual respect and recognition. The suspicion[65] that biological contingency "proves to be—in the very moment we can master it—a necessary presupposition for being-able-to-be-oneself and for the fundamentally egalitarian nature of our interpersonal

[60] Habermas (2003, 35).

[61] Ibid., 42.

[62] Cf. ibid., 25.

[63] Tömmel (2020, 183).

[64] Guignon (2004, 161) (original emphasis); cf. also ibid., 163: "So authenticity is a personal undertaking insofar as it entails personal integrity and responsibility for self. But it also has a social dimension insofar as it brings with it a sense of belongingness and indebtedness to the wider social context that makes it possible."

[65] I guess there is nothing more to it since Habermas himself concedes that "[q]uite literally, this essay is an *attempt*, seeking to attain more transparence for a rather mixed-up set of intuitions" (Habermas 2003, 22; original emphasis); cf. also ibid., 42: "Knowledge of one's own genome being programmed might prove to be disruptive, I *suspect*, for our assumption that we exist as a body or, so to speak, 'are' our body" (own emphasis).

relationships"[66] does not appear to be founded in this very biological contingency, but rather in a kind of essentialist prejudice.

As for the possibility of mutual relationships of recognition as free and equal persons, to date, the overwhelming majority of human beings admittedly (still) share the natural contingency of their genomes; however, this empirical fact is neither a necessary nor a sufficient condition for the *normative* assumption of *moral equality*. In fact, the (self-)ascription and mutual recognition of equal moral standing is less dependent on contingent biological underpinnings (except for determining the adequate kind of morality that human beings are in need of and equipped to endorse), but rather on an (equally contingent) historical and social process of experience, ethical reflection and (hopefully) learning. After all, the boundaries of the moral community are a matter of ethical argument and agreement, not of biological constitution.[67]

6.5 The Habermasean Argument in the Case of "Nana" and "Lulu" Revisited

To conclude, there is indeed hope that "Nana" and "Lulu" and other children like them will grow up to be autonomous persons with a healthy self-understanding of being authentic co-authors of their life histories and respected members of the moral community, irrespective of their genomes having been manipulated. A necessary condition for them to develop in this beneficial way is their being embedded in "relations of mutual recognition of communicatively acting persons" who "identify them as ones of us" and provide them with the capacity to identify themselves "simultaneously as a person in general, as a part or a member of [their] social community (or communities), and as an individual who is unmistakably unique and morally nonexchangeable",[68] instead of giving them a sense of being scientific guinea pigs or deplorably objectified beings deprived of true humanity. The key to rejecting the Habermasean argument lies in Habermas' own conceptual conditions for the development of personal identity, authenticity and autonomy.

References

Baker, Beth. 2016. The Ethics of Changing the Human Genome. *BioScience* 66: 267–273. https://doi.org/10.1093/biosci/biw017.

Baltimore, David, Paul Berg, Michael Botchan, Dana Carroll, R. Alta Charo, George Church, Jacob E. Corn, George Q. Daley, Jennifer A. Doudna, Marsha Fenner, Henry T. Greely, Martin Jinek,

---

[66] Habermas (2003, 13).

[67] Notably, humans share both biological contingency and 98.7% of their genes with chimpanzees; still, chimpanzees are not unanimously regarded to be members of the moral community.

[68] Habermas (2003, 35).

G. Steven Martin, Edward Penhoet, Jennifer Puck, Samuel H. Sternberg, Jonathan S. Weissman, and Keith R. Yamamoto. 2015. A Prudent Path Forward for Genomic Engineering and Germline Gene Modification. *Science* 348 (6230): 36–38. https://doi.org/10.1126/science.aab1028.

Baumann, Martina. 2016. CRISPR/Cas9—New and Old Ethical Issues Arising from a Revolutionary Technology. *Nanoethics* 10: 139–159. https://doi.org/10.1007/s11569-016-0259-0.

Cyranoski, David, and Sara Reardon. 2015. Embryo Editing Sparks Epic Debate. *Nature* 520: 593–594.

Doudna, Jennifer A., and Charpentier, Emanuelle. 2014. The New Frontier of Genome Engineering with CRISPR-Cas9. *Science* 346 (6213): 1258096. https://doi.org/10.1126/science.1258096.

Dzau, Victor J., Marcia McNutt, and Chunli Bai. 2018. Wake-Up Call from Hong Kong. *Science* 362 (6420): 1215. https://doi.org/10.1126/science.aaw3127.

Guignon, Charles. 2004. *On Being Authentic*. New York: Routledge.

Gyngell, Christopher, and Thomas Douglas. 2015. Stocking the Genetic Supermarket: Reproductive Genetic Technologies and Collective Action Problems. *Bioethics* 29 (4): 241–250.

Habermas, Jürgen. 2003. *The Future of Human Nature*. Cambridge: Polity Press.

Habermas, Jürgen. 2005. *Die Zukunft der menschlichen Natur. Auf dem Weg zu einer liberalen Eugenik?* Extended version, Frankfurt am Main: Suhrkamp.

Hildt, Elisabeth. 2016. Human Germline Interventions–Think First. *Frontiers in Genetics* 7, 81. https://doi.org/10.3389/fgene.2016.00081.

Kipke, Roland, Markus Rothhaar, and Martin Hähnel. 2017. Contra: Soll das sogenannte ‚Gene Editing' mittels CRISPR/Cas9 Technologie an menschlichen Embryonen erforscht werden? *Ethik in der Medizin* 29: 249–252. https://doi.org/10.1007/s00481-017-0435-y.

Lander, Eric. 2015. Brave New Genome. *The New England Journal of Medicine* 373: 5–8.

Lander, Eric, Françoise Baylis, Zhang Feng, Emanuelle Charpentier, Paul Berg, Bourgain Catherine, Bärbel Friedrich, J. Keith Joung, Jinsong Li, David Liu, Luigi Naldini, Jing-Bao Nie, Renzong Qiu, Bettina Schöne-Seifert, Feng Shao, Sharon Terry, Wensheng Wei, and Ernst-Ludwig Winnacker. 2019. Adopt a Moratorium on Heritable Genome Editing. *Nature* 567 (7747): 165–168. https://doi.org/10.1038/d41586-019-00726-5.

Lanphier, Edward, Fyodor Urnov, Sarah Ehlen Haecker, Michael Werner, and Joanna Smolenski. 2015. Don't Edit the Human Germ Line. *Nature* 519: 410–411. https://doi.org/10.1038/519410a.

Ledford, Heidi. 2015. The Landscape for Human Genome Editing. *Nature* 526: 310–311.

Liang, Puping, Yanwen Xu, Xiya Zhang, Chenhui Ding, Rui Huang, Zhen Zhang, Jie Lv, Xiaowei Xie, Yuxi Chen, Yujing Li, Ying Sun, Yaofu Bai, Zhou Songyang, Wenbin Ma, Canquan Zhou, and Junjiu Huang. 2015. CRISPR/Cas9-Mediated Gene Editing in Human Tripronuclear Zygotes. *Protein and Cell* 6 (5): 363–372. https://doi.org/10.1007/s13238-015-0153-5.

Nozick, Robert. 1974. *Anarchy, State and Utopia*. New York: Basic Books.

Pollack, Robert. 2015. Eugenics Lurk in the Shadow of CRISPR. *Science* 348 (6237): 871.

President's Council on Bioethics. 2003. *Beyond Therapy. Biotechnology and the Pursuit of Happiness*. New York: Dana Press

Pugh, Jonathan. 2015. Autonomy, Natality and Freedom: A Liberal Re-examination of Habermas in the Enhancement Debate. *Bioethics* 29 (3): 145–152.

Quante, Michael. 2018. *Pragmatistic Anthropology*. Paderborn: Mentis.

Regalado, Antonio. 2018. EXCLUSIVE: Chinese Scientists are Creating CRISPR Babies. In *MIT Technology Review* (November 25, 2018).

Regalado, Antonio. 2019. The DIY Designer Baby Project Funded with Bitcoin. In: *MIT Technology Review* (February 1, 2019).

Savulescu, Julian. 2001. Procreative Beneficence: Why We Should Select the Best Children. *Bioethics* 15 (5–6): 413–426.

Savulescu, Julian, and Guy Kahane. 2009. The Moral Obligation to Create Children With the Best Chance of the Best Life. *Bioethics* 23 (5): 274–290.

Savulescu, Julian, Jonathan Pugh, Thomas Douglas, and Christopher Gyngell. 2015. The Moral Imperative to Continue Gene Editing Research on Human Embryos. *Protein and Cell* 6 (7): 476–479. https://doi.org/10.1007/s13238-015-0184-y.

Segers, Seppe, Guido Pennings, Wybo Dondorp, Guido de Wert, and Heidi Mertes. 2019. In Vitro Gametogenesis and the Creation of 'Designer Babies'. *Cambridge Quarterly of Healthcare Ethics* 28: 499–508. https://doi.org/10.1017/s0963180119000422.

Tömmel, Tatjana. 2020. Reproductive Medicine and Parental Responsibility. In *Technology, Anthropology, and Dimensions of Responsibility*, ed. Birgit Beck and Michael Kühler, 177–191. Stuttgart: Metzler.

Birgit Beck is Assistant Professor (Juniorprofessorin) and Head of Department for Ethics and Philosophy of Technology at TU Berlin, Institute of History and Philosophy of Science, Technology, and Literature. Her current research interests include ethical and societal implications of (bio-)technological innovations. Her recent publications include *Technology, Anthropology, and Dimensions of Responsibility*. Techno:Phil—Aktuelle Herausforderungen der Technikphilosophie, Vol. 1, ed. by B. Beck, B. Gransche, J.-H. Heinrichs, J. Loh, Stuttgart: J.B. Metzler, 2020, co-edited with Michael Kühler; "Reproduktive Freiheit für wen? Einige Nachfragen zur Geltungsreichweite und Anwendbarkeit der Prinzipien der reproduktiven Autonomie und der wohltätigen Fortpflanzung," in: *Zeitschrift für Ethik und Moralphilosophie* 3, 2020, 127–135; *Kursbuch Bioethik*. Berlin: TU Universitätsverlag, 2019, co-edited with Jens Kurreck.

Chapter 7
Remainders of the Self: Consciousness as a Problem for Neuroethics

Marco Stier

Abstract Neuroscience has made great progress in recent years in detecting mental processes using electrophysiological and imaging techniques. One of the major breakthroughs in this regard has been the detection of residual consciousness in persons diagnosed with a vegetative state. However, this very success gives rise to a number of theoretically as well as ethically crucial follow-up questions. What neuroimaging can provide are highly artificial, mathematically processed images and not direct information about the patient's mind. Therefore, it is far from clear whether and to what degree behaviourally unresponsive patients are indeed conscious. This is all the more so since we are confronted with a double mediation: one between the person's mind and the image of the brain and the second between the image and something that we interpret as a meaningful answer of the patient to a question of the physician. For the patient in question it is of vital importance that his ability for consciousness is neither over- nor underrated since both can lead to significant suffering. Above that, residual consciousness or remainders of the person's self—even if reduced and fragmented—have to be taken into account for treatment or end of life decisions. Unfortunately, there is no consensus in neuroscience about what "consciousness" and "being conscious" really means. To be sure, there is a long tradition regarding notions like these in philosophy. However, philosophical insight does not find its way into neuroscience easily. Against this background, the paper aims at connecting philosophical and neuroscientific understandings of consciousness in order to better understand what may go on inside an otherwise unresponsive patient. The notion of a "self" that might be preserved even if the person in question is not fully conscious or only has some basic form of consciousness will serve as a working concept for this problem. Neuroethics has to solve the conceptual problem of consciousness in order to be able to solve the clinical problem of consciousness and with it the ethical problem of autonomy and the self.

M. Stier (✉)
Institute for Medical Ethics, History and Philosophy of Medicine, University of Münster, Münster, Germany
e-mail: marco.stier@ukmuenster.de

M. Kühler and V. L. Mitrović (eds.), *Theories of the Self and Autonomy in Medical Ethics*, The International Library of Bioethics 83,
https://doi.org/10.1007/978-3-030-56703-3_7

7.1 Introduction

What is the nature of consciousness? This is one of those perennial questions of philosophy that we are still unable to answer precisely but cannot neglect either. It has not made things easier that the neurosciences, too, have been occupied with this phenomenon for some decades. Rather, the problem is more pressing than ever due to the potential of today's emergency medicine and intensive care. On the one hand, we have various philosophical theories of mind and brain that often are neither compatible with each other nor with neuroscience. On the other hand, there is a daily flood of articles on "consciousness" in the neuroscientific literature in which this subject itself comes without any proper definition. At best, authors refer to William James who declared more than a century ago that consciousness mainly comprises awareness of the self and of the environment (cp. The Multi-Society Task Force on PVS 1994, 1501).[1] Given this lack of conceptual clarity of and medical knowledge about consciousness, together with the phenomenal inaccessibility of other minds, *neuroethics* should at least come up with a working definition of consciousness. Doubtlessly, neuroscience does have made great progress in detecting even minimal mental processes in recent years, using electrophysiological and imaging techniques. This very success, however, is one of two main roots of the ethical challenge. The other root is that medicine has become very successful in saving the lives of patients with traumatic or hypoxic brain injuries who would still have died a couple of years ago.

The ethical problem that arises from the combination of these two successes is the following: The life of those patients can often only be saved at the cost of a severe disturbance of consciousness. At least it seems so at the bedside. But what if there is still some form or degree of conscious experience left "deep inside" the person who is completely unresponsive? Is it possible to "detect" a patient's consciousness without communicating with him in some form or another? Is a communication conceivable—to say nothing of a real conversation—which takes place with the help of a MRI-scanner as an interpreter?

One of the major breakthroughs in this regard was the detection of residual consciousness in persons diagnosed as being in a vegetative state. When Adrian Owen and colleagues in a seminal paper claimed to have shown "beyond any doubt" that a patient with this diagnosis actually "was consciously aware of herself and her surroundings" (Owen et al. 2006) this triggered a flood of papers on clinical and ethical implications of these results. Yet, in all the years the neurosciences have not been able to reach a consensus regarding the meaning of the term "consciousness".[2] Up to the present day, many papers on disorders of consciousness not even contain a fragment of a definition this subject. Moreover, even if we had an uncontested and precise definition, what moral status does consciousness have? What status does *which degree* of consciousness have (Kahane and Savulscu 2009; Levy 2014)? And

[1] It is obvious that this understanding of consciousness is too simplistic. However, that is not the fault of James, whose account is richer, but the fault of those present authors who leave it at that.

[2] This is impressively shown by Płonka (2015).

if a patient with a brain injury does not have what we would count as full-fledged conscious experience—do the cognitive functions that may be left entitle us to speak of an "unconscious self" that has to be taken into account in ethical judgments in the clinic? Before the ethicist is able to make a moral judgment regarding consciousness he has to know what "consciousness" actually means.

One may argue that we only need to define the concept in theory to apply it in the clinic afterwards.[3] However, this would be a clear mistake. The case of patients with a disorder of consciousness gives rise to the assumption that the term "consciousness" alone is neither theoretically nor ethically the best landmark for neuroethics. For reasons that will become clear below, I propose to use the concept of "self" instead. Even if the self is no less contested in philosophy then consciousness, its advantage is nevertheless that it is possible to speak of unconscious parts of the self, whereas an "unconscious consciousness" would be pure nonsense. The self has room for hidden cognitive and/or emotional functions to be of moral value, what would be much harder to explain if we would orient ourselves exclusively on conscious processes and properties. Moreover, while *consciousness* is—in a sense—an all-or-nothing concept, it is possible to talk of "parts" or "remainders" of the *self*. A mental function or content that ceases to be conscious is usually regarded as having lost most of its moral weight: What has disappeared from the surface of consciousness has vanished altogether. Things are different if we think of the self as a person's basic functional fundament. The self can have remainders that are not conscious, but that we may have to take into moral account, nevertheless.

In the following, I will first turn to those various clinical contexts that may create the ethical problems I have just referred to. Thus, I will briefly discuss various *syndromes*, their nosologic development and their current understanding by neuroscientists. In particular, I will look into neuroscience's attempts to learn about possible conscious or cognitive functions that may be intact in a patient despite his unresponsiveness (1). The achievements of today's neuromedicine notwithstanding, the *limitations* of the methods applied are of particular importance here. In the next chapter, three basic neuroscientific accounts of the structure of consciousness are sketched (2). This is necessary for roughly understanding why one's underlying conception of consciousness plays a central role in research on minimal and degraded forms of consciousness. Both, the underlying conceptions and the applied methods can lead to fallacious conclusions that might even get dangerous for patients. In part (3) I will present five such conclusions illustrating why the para-clinical procedures alone are in a sense "blind". However, even if we can avoid such fallacies it should be obvious that neuroscientific results alone say nothing about how to treat and how to care for patients. What is needed in addition is an idea of the moral value of consciousness (4). This is all the more necessary as the patients in question are suspected to have only limited capacities for conscious perceptions, intentions and interests: What is the

[3]If this seems silly to you, it would not be much better the other way around: It is difficult—to say the least—searching for something in the brain that you do not know. Searching for a wolf in the woods presupposes to know what a wolf is.

moral difference between an undamaged and fragmented form of consciousness? In consequence of all these challenges, I propose to use the idea of a self and its possible remainders as a better landmark in neuroethics (5).

7.2 A Short History of Consciousness Disorders and Consciousness Detection

Thinking of "disorders of consciousness" (DOC) the first that comes to mind is probably the *"vegetative state" (VS)*, but there are more conditions falling under this category. Due to some form of brain injury a person may fall into a *"coma"*, a state in which she is neither awake nor aware of anything. Later, she may become awake[4] again, but remain completely unaware of herself and her environment. This is what we traditionally refer to as VS. If one holds that different syndromes of DOC are lying on lower or higher "levels",[5] systematically the next level to which a patient may arise is that of a *"minimally conscious state"* (MCS), followed by an *"confusional syndrome"* (CS). Thus, the main syndromes making up the group of "disorders of consciousness" are (1) Coma, (2) VS, (3) MCS, and (4) CS (Bender et al. 2015). A further condition often mentioned in this context is the *"locked-in syndrome"* (LIS), but since this is characterized not by disturbed consciousness but by the inability to move, LIS actually does not belong to this category.[6]

Although the VS-condition as such had long been known, it was not before 1972 that Bryan Jennett and Fred Plum coined the term "persistent vegetative state" (Jennett and Plum 1972). Other names that have been and partly are still in use are "apallic syndrome", or "coma vigile". In recent years, the label "unresponsive wakefulness syndrome" (UWS) has been introduced—not least to avoid the pejorative connotation of a "vegetable" (Laureys et al. 2010). Patients with this condition seem to be awake; they can open their eyes, they usually have an intact sleep–wake cycle, and can breathe autonomously. Even more complex reflexes can still be intact, such as smiling, erections or yawning. Still, these patients do not show any conscious reactions at the bedside. They cannot react in a consistent manner to spoken words nor follow a moving finger with their eyes. In short: people with this condition are not conscious—at least according to dominant definition.

After Jennett's and Plum's paper much remained unclear in the medical community. In particular, there was a need for a better understanding and description of the chances for recovery from unconsciousness. For this reason, in 1994 (after three

[4]Coma is usually a transient state from which patients either decline to brain death or improve to at least a vegetative state.

[5]Think of a multi-storey building: The basement represents coma, the first floor is the vegetative state, and so on up to the penthouse where the philosopher's second order self-consciousness resides. The level-based account has recently been challenged, see below.

[6]LIS does not belong to the DOC-family but it is nevertheless of utmost importance in the DOC-*Context* since a patient may be misdiagnosed with VS even though she is completely conscious (an impressive example give Vanhaudenhuyse et al. 2018).

years of working) a task force formulated criteria for distinguishing the "permanent" from the "persistent" vegetative state. Accordingly, the condition is to be called *permanent* if the patient's chance to regain consciousness is "exceedingly small", whereas the adjective "persistent" leaves open, which course the state will take in the future. Even though there are some reports of patients who regained consciousness even after several years, the vegetative state (resp. UWS) is now seen as permanent (and not only persistent) when it still prevails twelve months after a traumatic, and three months after a nontraumatic (e.g. hypoxic) brain injury (The Multi-Society Task Force on PVS 1994). Recently, Joseph Fins and James Bernat proposed to rename the "permanent" to "chronic" vegetative state because of the reports of late improvements (Fins and Bernat 2018).

Another major step in differentiating the various manifestations of DOCs was the introduction of the phrase "minimally responsive patients" by Joseph Giacino and colleagues (Giacino et al. 1991). From that on it took ten years of interdisciplinary discussion (see Fins 2015, chap. 8 for details) until the scientific community finally agreed upon the new *category* "minimally conscious state" (MCS) (Giacino et al. 2002). Patients in the MCS show behavior that can be regarded as a sign of consciousness but they remain unable to reproduce it consistently. Such a patient may follow simple commands, give gestural or verbal yes–no responses, and follow a visual stimulus with her eyes (Giacino et al. 2002, 350f). These and other signs of conscious awareness do not *each* have to be present in a patient to justify the diagnosis of a MCS. In some patients it is only one of them. However, phenomena are not as clear-cut in reality as in nosology. The difference between a patient being in UWS and the same patient having recovered to a lower state of MCS may in fact be minimal, whereas she may be aware of much more if she is in a later, more conscious state of MCS. This brought Marie-Aurélie Bruno and colleagues to subcategorize this syndrome into a lower level form and a higher one: MCS^- and MCS^+ (Bruno et al. 2011).

It can be challenging to make the correct diagnosis for a patient who seems to be awake but does not respond to external stimuli or does not so appropriately. This is illustrated by the disturbing discovery that up to 40% of those diagnosed with a UWS in fact are at least in the MCS^-, that is, they are conscious to a certain degree but cannot communicate with their environment (Schnakers et al. 2009).[7] Since "the absence of evidence for awareness is not tantamount to evidence for the absence of awareness" (Jox and Kuehlmeyer 2013, 1) researchers tried to find out by way of technological means whether or not a given patient is actually aware of something. Four paradigms of doing this have been tried out during the past twenty years.

In the *passive paradigms,* patients receive certain stimuli (usually auditory or visual) while their specific neurophysiological responses are detected by neuroimaging or electrophysiological methods. If the patient's brain reacts differently to, e.g., familiar and unfamiliar voices (de Jong et al. 1997), she must be aware of the difference, so the idea. Similar studies have been undertaken using pictures of faces

[7] Again, this must not be confused with LIS, where the patient is perfectly clear but cannot communicate.

(Menon et al. 1998) or the patient's own name (Staffen et al. 2006).[8] Nevertheless, there was some agreement that one should not simply infer consciousness from mere brain activity. We all can do amazingly complex things while being absent-minded. For this reason, researchers proceeded to so-called active paradigms.

Active paradigms require the patient to cooperate in the diagnostic process. Here, the mere fact that someone cooperates and "reacts" properly seems to ensure that he is indeed consciously aware of what he is asked. To react in accordance to an instruction *presupposes* that the patient is consciously aware of the very instruction. In contrast, the mere activity of, say, the auditory areas of the brain cannot prove that the patient has any conscious experience of what is going on. This was the idea of Owen and colleagues (Owen et al. 2006). The researchers asked patients laying in the MRI scanner to imagine playing tennis or walking through her home. The patients' brain scans turned out similar to those of healthy controls performing the same mental tasks of imagining tennis playing or walking. Apart from the fact that an experiment like this will only work if the patient has an intact hearing capacity and is able to understand verbal language, the inference to consciousness—let alone *self*-consciousness—is somewhat daring. Just to cite two objections: It is well possible that the responses from the patient in the imagery task is only the unconscious reaction of the brain to the last word of the investigator's command ("home" or "tennis") (Greenberg 2007). Secondly, while a certain behavior or thought necessarily implies a related brain activity, the reverse is not the case (Nachev and Husain 2007). Indeed, the latter is something that philosophers know for decades.

Communicative paradigms: The next step in the quest for residual consciousness was to establish a basic form of communication. A patient who was diagnosed with UWS for five years was instructed to use either the motor imagery (playing tennis) or the spatial imagery (going through her home) to give "yes" or "no" answers. In this way, it was possible to get correct answers to biographical questions from the patient (Monti et al. 2010). Davinia Fernández-Espejo and Owen reported a similar case. Their patient had been considered to be a UWS-patient for 12 years before they finally could contact him using this approach (Fernández-Espejo and Owen 2013). While there are many reports in the literature about UWS/MCS patients whose brains showed some reactions to stimuli (the passive paradigm), cases of obvious command-following (active paradigm) are fewer. Reports of successful technology-based communication are still anecdotic.

Network paradigm: As already mentioned, there may be patients for whom none of the above paradigms produces satisfying results in the search for covert consciousness. If one does not *find* awareness in a non-responding patient, this does not need to mean that there *is* none. She may just lack the adequate hearing or verbal capacities. If this is the case and the patient is in principle still able to have conscious experiences, one can test the functional integrity of those neuronal networks that are a prerequisite for awareness. In particular, two such networks have been proposed (Vanhaudenhuyse et al. 2011): a "default mode network" (DMN) and an "attentional"

[8] I can only give some hints to research literature and methods here; see Cavaliere et al. (2016) for a concise overview.

or "external awareness network" (EAN). Since Raichle et al. (2001) showed the existence of a neuronal network that is more active when the brain is at rest, i.e. when it is not engaged with a specific task (the DMN), much research has been done on this subject. In particular, it is now being assumed, that the DMN is involved in the processing of the self (recall of autobiographic memory for instance), self-centered thoughts, self-reference, or prospective personal projection (see Mäki-Marttunen et al. 2016). EAN, in contrast, is active during externally oriented tasks, such as the cognitive selection of sensory information (Corbetta and Shulman 2002). In short, DMN goes to the inside, EAN to the outside.[9] If the functional dynamics between these two networks deteriorates, awareness may be affected or the experience of the environment may become "kaleidoscopic and chaotic" (Naro et al. 2017, 166). If a patient is unresponsive, and the passive or active para-clinical examinations do not have results either, then the preservation of DMN and EAN could at least give a hint that there is something left "inside", maybe even (some form of) consciousness.

Researchers fight for ending or at least relieving potential silent suffering of DOC patients. There is a danger, however. According to Immanuel Kant, not only "[are] thoughts without content […] empty", but also "[are] intuitions without concepts […] blind" (Kant 1998, A51/B75). Scientific intuitions are sometimes as blind without the philosopher's conceptual armamentarium as philosophical ideas about consciousness occasionally tend to be empty without the content provided by scientists. The resulting danger for the patient is that we might ascribe consciousness where in fact there is none[10] or else that we negate the presence of conscious experience where someone in fact has enough awareness to be able to suffer. So, what do neuroscientists think about consciousness?

7.3 Neuroscientific Accounts of Consciousness: Levels and Dimensions

We have learned above that there are different variations of DOCs and many attempts to detect (conscious) mental processes in the brain of unresponsive patients. Now we need at least some idea of what consciousness is in the eyes of neuroscience. It is tempting to assume that today's neurosciences, equipped with sophisticated methods and machines, have a clearer and more reliable account of consciousness than philosophers, who are only equipped with a bunch of more or less old-fashioned ideas. The latter, then, would be primarily speculative while the former would be based on exact measurable result and hard facts. This picture is inadequate, however, since there is

[9]It goes without saying, that things actually are much more complicated.

[10]One could argue that an unconscious patient cannot be harmed because in this case "*there is* nobody" who could be harmed. However, harm can be more direct (e.g., pain) or more indirect (e.g., violation of interests). Moreover, one has probably to take into consideration various dimensions of consciousness and the self (for more on the concept of self see part 5 below).

no such thing as a "consciousness meter". Before a researcher can examine an unresponsive patient's "consciousness" in the scanner he has to develop some idea, i.e. at least some provisional working-concept of what consciousness could be. No device whatsoever will be able to detect consciousness without a researcher's prior conceptual work. This is because unless you know beforehand what you are searching for you will not find anything.

At first sight, it seems self-evident that consciousness is not an all-or-nothing phenomenon. Sometimes you are drowsy when you have just woken up from deep sleep, sometimes you are so focused on a task that you barely notice anything around you, and some of us know how consciousness feels if one has had a beer too many. Sometimes, of course, you are so alert that you can balance four ideas in mind while writing down the fifth. Yet, that does not say much about the inner structure of consciousness.

It is contested whether it is possible to distinguish *levels* or degrees of consciousness or whether we should better talk of *dimensions*. The answer to this question is important, since all the traditional diagnostic categories introduced above, from coma to full consciousness, rely on the idea of levels (Posner et al. 1983). One drawback of this model is that it is far too vague. Even if one uses the revised Coma Recovery Scale (CRS-R) with its scale ranging from zero to 23 points, one is dependent on the behavioral response of the patient at the bedside. This does not necessarily mean the CRS-R were wrong, it is—like the other coma scales, too—simply not helpful for the detection of *covert* consciousness. Another and deeper problem is that the rather "simple" one-dimensional levels model of consciousness could itself be wrong. May the coma scales be as reliable as possible; the idea that consciousness shows degrees and levels could nevertheless be misleading.

This is what Tim Bayne and colleagues have stated (Bayne et al. 2016). In their view, global states of consciousness are better understood in a multidimensional model that does without levels. Global states of consciousness, they say, "characterise an organism's overall conscious condition"; and since it is the *overall* condition, an organism can be "in only one global state of consciousness at a time". Local states, in contrast, "are individuated in terms of their contents or phenomenal character" (Bayne et al. 2016, 406). The authors argue that at present theorists wrongly assume that the different disorders of consciousness correspond to different, specific "levels" of global states of consciousness. Consequently, UWS would correspond to one level, MCS^- to another, higher one, MCS^+ to a third, and so forth. What the authors have in mind, then, seems to be a "horizontal" distribution of the dimensions of consciousness rather than a "vertical" graduation of the consciousness abilities of a patient altogether.[11]

The level-based understanding of consciousness—in contrast—suggests that one individual can have a higher degree of overall-consciousness than another one. In other words, (to use the building analogy once more): While person P2 lives on the second floor, P1 is on the first floor or in the basement. For instance, a person in REM-Sleep and another person in mild sedation would have to be on different levels

[11] Actually, consciousness corresponds to a single-storey building in this picture.

of consciousness, the one being absolutely "higher" than the other. According to Bayne and colleagues, it is not the *grades* of consciousness but the *dimensions* of cognitive function and experience that are different. P1 and P2 are not on different floors, so to say; they rather are on the same floor, but in "differently furnished" apartments, one well equipped with all comfort, the other only having a table and a chair in it. Their first and main critique on the level-account is that consciousness is not a gradual but an all-or-nothing phenomenon, because it depends on whether an individual possesses a subjective point of view. Since subjectivity is not gradable and since together with subjectivity consciousness comes into existence, consciousness is not gradable. Either you are pregnant—or conscious of something—or you are not, but you can be neither "a little pregnant" nor "a little conscious". A second argument against the level-account is that global states of consciousness allegedly cannot be assigned "a determinate ordering relative to each other" (Bayne et al. 2016, 407). Different global states of consciousness are not "higher" or "lower" in relation to each other; they are merely different regarding their *dimensions* and not more or less conscious. To borrow a phrase from Bayne and colleagues: One patient is probably not "more conscious" than another, but "conscious of more". This "being conscious of more" does not simply mean that I can discern seven birds in the tree and you only four. States of consciousness probably have numerous dimensions and relations between these dimensions that are still largely unknown and a matter of hypotheses.

Another account that is informative for the purpose of this essay takes up the multi-dimensional view but also keeps the idea of graded states and levels of consciousness (Fazekas and Overgaard 2018). The model of consciousness just sketched refers to the ideas such as fragmentation and availability of conscious content. Peter Fazekas and Morten Overgaard incorporate these elements in their multi-factor model. The authors keep consciousness-degrading factors like "fragmentation", but they extend this idea to a complex three-dimensional conception of conscious awareness. Even if their model is still speculative to a certain degree, it is nevertheless helpful for getting a better grip on the ethical challenges of DOCs. In light of the increasing complexity of the question, "Is the patient conscious?" we should realize that "to be conscious" is at best the first brick of a more fundamental answer to the questions how to treat patients in states of covert and/or possibly fragmented and faded states of consciousness. At least, much will be gained if neuroethicists get a clearer picture of what about consciousness is considered valuable and what possibly not. The main difference of multi-factor model to the one of Bayne and colleagues is that according to the latter consciousness is an all or nothing phenomenon while the former allows for degrees. In other words, one person may not only be "conscious of more" than another, she may be "more conscious" as well. More precisely: P2 can be "more conscious of more" as well as "less conscious of less" than P1 (and maybe even more conscious of less).[12]

According to Fazekas and Overgaard, four main factors determine the conscious perception of a stimulus: (i) the *level of processing* of incoming perceptual information, (ii) the *intensity,* (iii) the *precision*, and (vi) the *temporal stability* of a neural

[12]Think of a patient who is fully conscious of her pain, but of nothing else.

representation. These four factors are responsible for the way in which a person is consciously aware of something. A brief sketch will explain why they are important for ethical judgments regarding DOCs. As it is the basic structure of conscious perceptions and not their exact neuronal encoding that is of interest here, I think we can take this as a model for other senses and—to a certain degree—for conscious contents in general.

(i) *level of processing*: The post-perceptual processing of an object can take place on different, hierarchically structured levels. For instance, the whole word "SCANNER" together with its meaning and connotations will be processed on a relatively high level, whereas on lower levels only the letters of the word or even fragments of the letters are represented. In this way, a perception can be fractured and fragmented.
(ii) *intensity*: This refers to how clear something is represented. Take a single letter of our word as an example. An intensive neural response will lead to a clear and salient perception of the letter, but it will be less clear and "faded" if the intensity is only low.
(iii) *precision*: Independent of the intensity a neural representation (and consequently the phenomenal perception) may vary in its precision. If the representation is less precise, the perception will be more generic or ambiguous. Think of our word "SCANNER" or of a single letter of it that may be sharp or "blurred" so to say. What one perceives as "something green", to use a different example, may actually be a tree, a bush or even a red bicycle.
(iv) *temporal stability*: The temporal stability of a neuronal representation determines whether there is a full conscious perception, only a very short one, or no perception at all. As Fazekas and Overgaard explain: "High temporal stability corresponds to stable, low temporal stability to flash-like conscious percepts." (Fazekas and Overgaard 2018, 1841)

Factors (ii) to (iv) can degrade a conscious perception at each of the levels of post-perceptual processing mentioned in (i).

Admittedly, this is only a rough and partial sketch of what is being under debate in cognitive science. However, my claim is that we should consider insights like these in our neuroethical judgements.

7.4 Some Fallacies

At this point, we can draw a first conclusion. It comes as no surprise, but it is nevertheless important for our ideas of the self and autonomy and for our neuroethical judgments: Consciousness can be "fragmented, faded and flickering". Let us call this the "triple-f problem" (TFP for short). It does not make things easier if we conceive of global states of consciousness as something with dimensions instead of levels: When someone is "clearly" conscious of something, but this something is but one dimension of many of a more complex object or context, then it would be misleading to say that she "is conscious" or "has consciousness".

On the background of the neuroscientific discussion, we can identify at least four possible epistemological fallacies[13] that can lead to misattributions of consciousness. We may attribute too little consciousness to the patient or too much or we may attribute the wrong kinds or (iv) contents of consciousness. Each case can have problematic, if not devastating consequences for the patient.

Local-global fallacy: Most of the experiments sketched above have at least shown that the brains of patients who formerly would have been regarded as being in a vegetative state actually can process certain kinds of information. From this one could infer certain *local* states of consciousness (but see the next fallacy). In line with TFP this does not yet imply a certain level or kind of *global* consciousness. It is not possible to infer the character of a global state of consciousness only from the presence of some local states. The same could possibly be true for the opposite direction: That we can ascribe "all in all" a certain level of *global* consciousness does not justify the ascription of certain kinds of *local* consciousness. An example for the latter would be a person being in a lower, somewhat faded global state of consciousness who cannot experience pain consciously or who has no conscious access to biographical information. The opposite may be true just as well. Therefore: it is neither justified to infer an overall, global state from a local state of consciousness nor to make an inference in the opposite direction (from global to *specific* local states).[14]

Neuronal-phenomenal fallacy: Georg Northoff has argued that patients with DOCs may present some "neuronal-phenomenal dissociation" (Northoff 2016, 110). Even if we detect specific stimulus-induced or task-related neuronal activity in a patient, this does not guarantee that she phenomenally experiences anything. This again is true in the opposite direction. The absence of *detectable* neuronal activity is not tantamount to the absence of all neuronal activity and does not justify strong claims about the absence of conscious experience (Laureys and Schiff 2012, 482f).[15] The fallacy does not pertain to the inference from neuronal activity to consciousness in general, but it pertains to the claim that this inference is *sufficient*. Therefore: It is not justified to infer (phenomenal) consciousness from neuronal activity alone.

Signal-meaning fallacy: The latter applies even more to *meanings*. Even if the patient in question consciously experiences something, we cannot know *what* precisely this is. Even if the researcher would see the word "bar" appearing on the screen while scanning his patient's brain, he could not tell whether this relates to "Bloody Mary", to "chocolate", "wood", or to nothing at all. Moreover, if it relates to "Bloody Mary", he could not be sure if it is a drink or a murderous woman. One of several possible reasons for this problem is that, due to TFP, elements of what a

---

[13]More potential fallacies have been described in regard of the broader ethical context, see for instance Johnson (2016), Northoff (2016).

[14]A patient may be familiar with his mother's voice or his father's name without knowing much about his own and his parent's biography.

[15]This is not to say that consciousness is possible without brain activity.

patient formerly knew might have become isolated from each other. Therefore: It is not justified to infer a specific meaning from neuronal activity.

Environment-self fallacy: A patient's brain may be able to process sensory input and the patient may be consciously aware of, say, spoken words, music, or pictures. Since it is assumed that two networks operate in the brain—the DMN for the processing of internal information and the EAN for "the outside" (see above)—both networks might become deteriorated and/or dissociated from each other. We can hypothesize that this could possibly result in various forms of informational disintegration. External stimuli might be experienced in the absence of conscious self-related information that are a basis of a comprehensive personal self. It might even be the case that there is still an "inner self", while stimuli coming from the outside are blocked or cannot be properly integrated anymore. Therefore: It is not (always) justified to infer from the neuronal processing of external stimuli that an "inner" personal self is still present.

These four epistemological fallacies lead us to another problem: a possible *ethical* fallacy. Currently, we cannot be sure whether an unresponsive patient is conscious or not.[16] If she is conscious, we cannot be sure *what* she is conscious of. And even if we can approximatively tell what she is conscious of, this still does not provide us with straight ethical orientations. What we thus need is a sound idea of the moral value of (certain sorts of) consciousness, because, e.g., it does not seem warranted to conclude just from the presence of some consciousness in a patient that keeping her alive is always in her best interest. Let us call this the *consciousness-value fallacy*. It goes like this: It is not justified to infer from the mere presence of some sign of consciousness alone that it is of moral value and the patient's life worth of protection. Before I discuss the "remainders of the self" in the next section, I have to dwell a bit on the moral value of consciousness.

7.5 The Moral Value of Consciousness

Joseph Fins rightly urges us to acknowledge adequate diagnostic measures and rehabilitation programs as a *civil right* for patients with DOC (Fins 2015). Therefore, the question is not whether such a patient should be helped if she could be helped. However, as is the case elsewhere inside and outside medical ethics, a right to something does not imply an obligation on its bearer. A right to life is no duty to life. That a patient has a right to receive a thorough diagnosis and the help she needs in her pursuit of recovery does not indicate by itself which procedure is morally adequate in the individual case. In particular, when it comes to patients with severe brain injury a question arises that might be overlooked in everyday clinical practice. That is the question of the "moral value of consciousness". In this paragraph I take for the sake of clarity only patients into account who have lost their capacity for (certain

[16] In principle, we can *never* be completely sure whether someone *really* is conscious. See the zombie arguments in the philosophy of mind.

kinds or levels of) consciousness irrevocably. Ethicists have warned from early on that it depends on the kind of consciousness detected which ethical implications follow regarding DOC. A minor increase of consciousness, say from UWS to MCS^-, could theoretically be celebrated as a therapeutic success. In the patient's subjective experience, however, it might amount to a catastrophe (Glannon 2016; Kahane and Savulscu 2009; Wilkinson, Kahane, and Savulescu 2008). For this reason, Laura Cabrera and Judy Illes suggest that the aim of interventions must be to restore self-determination, communication and the ability to engage with the environment and not only consciousness (Cabrera and Illes 2018).

Most ethicists will probably agree that consciousness matters morally in some way. But what exactly makes conscious thoughts and feelings valuable? In particular: what is the moral significance of the difference between the full-fledged conscious mental life of a healthy human being and the minimal presence of some form of residual consciousness (Wilkinson et al. 2008, 31)? A central idea is that it matters if the patient can suffer pain. Pain, in turn is at least conceptually connected to what philosophers call "phenomenal consciousness". Hence, phenomenal consciousness is what matters most in ethical regard. This would be a quick and easy answer, if only we knew what phenomenal consciousness really is and how (and whether) it comes always in tandem with what has been called "access consciousness", "sapience" or "informational consciousness". The term "access consciousness" was coined by Ned Block (1995) and is still of much influence. It is the type of conscious content that can be used "to control reason and behavior", as Block puts it (Block 1995, 229). Since patients with DOC usually cannot "behave" in a strict sense, Guy Kahane and Julian Savulescu prefer using the term "sapience". Sapience "involves a rich mentality that is likely to imply consciousness in another, nonphenomenal sense" (Kahane and Savulscu 2009, 11).[17] For the same reason Will Davies and Neil Levy speak of "informational" consciousness (Davies and Levy 2016, 130). *Phenomenal* consciousness on the other hand refers to Thomas Nagel's famous "What is it like?" (Nagel 1974). It "feels" somehow to see a red apple, to smell your partner or a pizza. It even "is" somehow to *be you.* Despite all the literature on phenomenal and access consciousness, what is of interest at this point is the following: one can experience a certain "feel", an "it is like *that*" on the one hand. On the other hand, there is the more explicit, declarative informational content of our mental life. It is still an open question whether both types of consciousness can come apart. Often enough phenomenal consciousness is used synonymously with consciousness *as such*: If you experience something consciously you will experience it *phenomenally* (Kahane and Savulscu 2009). I will refrain from taking a stand in this debate here. In light of our gaps in knowledge regarding the neuroscience of consciousness it is nevertheless reasonable assuming that both can indeed come apart. Thus, a patient may (a) be phenomenally conscious without having access consciousness (let alone *sapience*), (b) have declarative informational content consciously available in the

[17]*Sapience* refers to Block's *access consciousness* but seems to have a more comprehensive meaning. While Block's concept can be used (locally) for the naked information, *sapience* alludes to the person's (global) mental life.

absence of an accompanying phenomenality, (c) have the usual kind of experiences, where information is (nearly always) coupled with a phenomenal "feel".

Let us first take a patient with phenomenal consciousness but no access consciousness. This patient is probably able to experience physical pain and other kinds of discomfort. She may even feel depressed, as far as a "nonverbal" kind of depression is possible. This depression may be of a sort that does not consist in a pessimistic self-referential outlook. Rather it would affect the overall mood of the person. Since this patient has no access consciousness, she will have no present critical interests,[18] no idea of her personal life and no notion of gains and losses. According to Levy and Savulescu, such a patient would not have a right to life. We would have the moral obligation to make sure that the patient feels comfortable, that she is not in pain; but in the irreversible absence of interests, of plans and of a concept of future, she could not have any interest in the continuation of her existence. However, phenomenal plus *access* consciousness is not enough for a right to life either. The reason is that "only a self-conscious being can have preferences regarding how its life goes" (Levy and Savulescu 2009, 368), and access consciousness as such does not necessarily pertain to the kind of higher order consciousness that is necessary for those preferences: namely *self-consciousness*.[19] This is already shown by the fact that (access) consciousness in various ways can become fragmented and flickering (TFP). I will come back to this latter case in a moment, since it is of importance for DOC.

In the second scenario, the person in question has full access consciousness and even something similar to self-consciousness. He is what philosophers call a "phenomenal zombie".[20] Imagine the zombie is your twin. He would be much like you, except that he has no phenomenal consciousness. That zombie-twin would know what you know[21] and he would behave the way you behave. The only difference between you and your zombie-twin is that you are able to understand the meaning of the "what is it like" phrase while your twin is not. He feels no pain, no sadness and no love—at least not in the way we non-zombies usually do. Whether it is possible for such a phenomenal zombie to have interests and a subjective point of view is disputed. According to Kahane and Savulescu—who even use "phenomenal consciousness" and "consciousness" interchangeably—a being without phenomenal consciousness can have neither interests nor a subjective perspective (Kahane and Savulscu 2009, 17). In contrast to that, Levy argues convincingly that subjectivity does not require phenomenal consciousness (Levy 2014, 132ff).[22] If he is right, then phenomenality is not everything that is of moral value in consciousness. Above that, it makes the thesis

[18]That critical interests can persist in dementia, unconsciousness and even death need not be disputed here.

[19]Remember that I assumed for the sake of clarity that the patient has no prospect of recovery.

[20]Block himself argued for at least the conceptual possibility of phenomenal zombies, i.e. creatures with access—but no phenomenal consciousness (Block 1995).

[21]Actually, he is not only a philosophical *zombie*, he is also a "*philosophical twin*". This philosophical twin is your exact duplicate *minus* phenomenal consciousness.

[22]Levy's respective passage is worth citing at length: "But even if it is true that only a conscious being has a point of view, and that there can only be values if there are conscious acts of valuing, I see no reason at all to believe that the consciousness in question has to be phenomenal consciousness.

intelligible that consciousness can indeed come in degrees. Namely, if the contents that are access-conscious are diminished or fragmented (see TFP above). Joshua Shepherd recently argued for several forms and spaces of phenomenal consciousness and its moral value. If phenomenal consciousness has moral value, it has it in virtue of experiences. And the phenomenal value of those experiences is embedded in and dependent on a sort of network that represents the subject's "evaluative space" (Shepherd 2018). Moreover, the possibility of consciousness that is *not* phenomenal consciousness has also implications for the AI debate. A robot that shows a sort of self-consciousness but that does not *feel* "what it is like" to have its battery charged or its joints lubricated may be easier to imagine than a weeping computer.

Two scenarios worth mentioning are left. The first is (c) mentioned above, it can be treated in passing. If a patient has the usual kind of experiences, meaning information is (nearly always) coupled with a phenomenal "feel", then he has the "usual" moral status, period. This is actually the case with patients suffering from a locked-in syndrome. The second scenario—an extension of (a) that I have not mentioned yet—is regarded by lay people as well as ethicists as probably the worst case: a patient in a DOC shows only scattered, diminished forms of phenomenal and access consciousness without genuine sapience (or self-consciousness). To this I will turn in the following Chapter.

7.6 Consciousness and Remainders of the Self

Let us imagine a patient can feel pain, and she even feels comfortable (or uneasy) when she hears a certain voice, even if she cannot identify it as her husband's voice. A number of "zombie systems" (see Levy and Savulescu 2009; similarly Monti 2012; Pistoia et al. 2016) may still be working with proper information, but this information is not available for self related, meta-representational functions.[23] That means the patient's brain processes certain inputs and is even able to connect this inputs to an existing biographical context, but all this happens on a sub-personal level, a level without conscious reflection. Due to TFP (see above) the brain—and, hence, the patient—cannot combine the remaining bits of information to a concept of self, to intentions, a life plan or an idea of her personal past or future. We may even

---

My zombie twin has a point of view. He sees the world from a particular perspective, in an attitude-infused way. Indeed, his idiosyncratic take on things is identical to mine. He also values things. He believes that his life will go better if certain things happen, and therefore he desires that those things happen. He is motivated to do things that make it more likely that those things happen. There are states of affairs that he does not believe he can influence, but toward which he takes attitudes too (he hopes for peace in the Middle East, for instance). He is inclined to say that he values these things, and I am inclined to agree with him." (Levy 2014, 133).

[23]In the philosophical literature on consciousness is the "dreaming driver" a well-known example from everyday experience: Once you know the route from home to your office by heart, you will be able to make the ride "automatically" while dreaming of your lover. It may be the same with the zombie-systems giving the investigator the wrong impression that the patient is conscious.

assume, for the sake of the question, that bits of biographical episodes or "fragments of the here-and-now" pop up into consciousness from time to time. Is this patient conscious? This is presumably the hardest of all problems of consciousness. It would be plainly wrong calling a person in a minimally conscious state unconscious; at the same time, it would be misleading to call her conscious in a usual sense. This is not only a problem "in" but also "for" neuroethics. Since we cannot wait for a proper concept of consciousness to come, it must suffice to take at least note of the problem and its ethical ramifications for the moment.

Is consciousness all that matters? It might seem so after what has been said so far particularly on its phenomenal form. Considering *first* our gaps in knowledge and *second* all the un- or preconscious functions of the human brain it might be advisable to be cautious.

This is the spirit of a paper by Michele Farisco and Kathinka Evers who argue for the ethical relevance of the unconscious (Farisco and Evers 2017). Neuroscience's insights of recent years show that the human brain is able to perform highly complex tasks without any awareness (the "zombie systems"). This is not only true for the healthy brain. The vegetative or even the comatose brain, too, can retain astonishing high-level cognitive functions (Eickhoff et al. 2008; Monti 2012). On this background, Farisco and Evers have two arguments to which I will add a third one in a moment. *First,* the authors state, "if consciousness is ethically relevant by virtue of certain characteristics, and the unconscious shares those characteristics, then the unconscious is ipso facto ethically relevant". *Second,* they present an argument that they themselves regard as weaker than the first one: Since the conscious is regarded ethically relevant, and since the unconscious and the conscious stand in a relation of complex mutual influence, the unconscious, too, is of ethical relevance (Farisco and Evers 2017). Given the newer results of neuroscience, it can be argued, according to the authors, that unconscious emotions are possible, and that it is not unreasonable to assume the existence of an unconscious well-being and of unconscious interests. In contrast to other authors who allow for subjectivity only under the condition of consciousness of some kind, they "conclude that not only awareness but also unawareness is something highly subjective, in the sense that it is developed in a peculiar and subjective way " (Farisco and Evers 2017, 4). And indeed, it is intuitively plausible that there could be an ethical difference between a being that is "unconscious, period!" and another being that is unconscious with residual cognitive functions and a preserved subjective (even if unconscious) focus. Of course, that does not mean that the unconscious is morally on a par with the conscious.

Before coming to some concluding remarks I will add a third argument in favor of the ethical relevance of the unconscious and partial conscious (the "fragmented, faded, and flickering consciousness", so to say). This argument refers to the aspect that gave this essay its title: the remainders of the self. If we take into account that consciousness is a thing that has many forms and functions, and if we accept (at least for a moment) that the pre- and unconscious is not without any moral value, than the idea of a person's self can give us some orientation in ethical considerations on disorders of consciousness.

Think of the self as an iceberg. Just as the main part of the iceberg is under water, the main part of the self is below the surface of consciousness. This is true at least with regard to what most of us would pretheoretically hold for the self, namely *self*-consciousness. Self-consciousness is only the tip of the iceberg. This essay is certainly not the place to debate all the many theories of and ideas about the self that are available on the philosophical market. Therefore, it must suffice to mention that most theories include lower and higher parts of the self. These parts differ regarding the level of consciousness they require, regarding their abstractness and regarding the kinds of content they integrate. One lower form, for instance, is Ulric Neisser's "ecological self" which represents a diffuse and implicit form of bodily awareness, a sort of a pre-conceptual here-and-now (Neisser 1988). This is probably in line with Antonio Damasio's "proto self" and "core consciousness" (Damasio 1999). Similarly, Kai Vogeley and Shaun Gallagher speak of a "minimal self" that entails "pre-reflective, implicit, or tacit aspects of our experience" (Vogeley and Gallagher 2011, 119). Recently, Michael Schaefer and Georg Northoff stress the importance of the unconscious "embodied" parts of the self for our conscious mental life (Schaefer and Northoff 2017). Neisser's theory of the self is a helpful tool to a better understanding of what may be at stake for patients with minimal consciousness. This holds even though his account does not need to be seen as the last word regarding the structure of the self. In any case, it can serve as a proxy for other theories, which differ only in detail, not in their core idea.[24]

Neisser distinguishes five kinds of self. The first and lowest is the aforementioned ecological self. A second one is the *interpersonal* self. According to Neisser, it is "the self as engaged in immediate unreflective social interaction with another person" (Neisser 1988, 41). These two kinds are rather low with respect to consciousness and abstraction. Things become more complex with the remaining three kinds: the *extended*, the *private*, and the *conceptual* self. The extended self reaches from the past (in the form of personal memories) into the future (in the form of anticipations). It may also be called an explicit biographical self. The private self focuses more on the hidden mental life of a person, on the fact that there are conscious experiences not available to others. It is obvious that the different kinds of self overlap logically as well as experientially. We usually do not have autobiographical memories and a private mental life on top of them. Both kinds of self and self-knowledge are interwoven most of the time. It is nevertheless conceivable that both can come apart. This would be the case when one has some sort of inner mental life but neither conscious biographical memories nor explicit expectations for the future (or even an idea of it). In this case, a person would be trapped in the present, so to say. The reverse case, too, might be possible in reality (and not just in a philosopher's imagination): here one would have retained biographical memories (which do not need to be of a sophisticated and high-level sort) but no full-fledged inner mental live. Take a photo album and a movie as an analogy: the preserved but isolated biographical memories would be the album whereas a genuine private mental life—containing and connecting these

[24]For an instructive comparison of different accounts of consciousness and the self see Morin (2006).

memories—would be more like a movie. The conceptual self, finally, is probably the one that most closely resembles the folk understanding of a self. Neisser mentions several important "theories" that are usually part of the conceptual self. The first are "role theories" that give us a notion of our place and function in society. Secondly, there are the "internal models" that comprise theories of our body and our mind. Neisser is somewhat vague about both. It is probably not entirely false if we say that our theory of our body gives us a biological model of ourselves while the theory of our mind—rather an object of (folk-)psychology—provides us with a notion of what a mind is and how it works.[25] This latter aspect of the conceptual self already points to the third one: "trait-attributions". If the conceptual self as such comes close to the folk understanding of what a self is, the trait-attributions come even closer. All of us have an idea of what we are and of how good or bad we function and manage life. "We may belief", Neisser explains, "that we are clever or stupid, handsome or ugly." (Neisser 1988, 54). Here narrative theories of the self come to mind. Despite the many narrative accounts of the self (see Schechtman 2011 for an introduction), we can probably take for granted that each person has a sort of what Daniel Dennett has called a „center of narrative gravity" (Dennett 1992). In contrast to Dennett, however, who considers the self a mere fiction, it is more adequate to treat it as something objective. This does not mean that the individual has a sort of "real self" that he can miss. A person's self is indeed a matter of narration and interpretation, but that does not make it a fiction in the way Albus Dumbledore is. A real person's self has a kind of causal influence in the world that Dumbledore's has not.

Now, what is the introduction of the self and its remainders into neuroethics good for? Consciousness is more than neuronal firing, and self-consciousness is more than just being consciously aware of a given stimulus. Between the extreme ends of coma and self-conscious life planning there is room for many complicated and fragmented forms of unconscious information processing, conscious perceptions, fragmented bits and lost islands of consciousness. Neuroscience only begins to decode the material basis of the human mind. Philosophers have to take neuroscientists serious and vice versa. Moreover, neuroethicists need to develop a concept of the human mind and the mind's functioning that can serve as a starting point for the current and future ethical challenges. One such challenge is the enigma of the many forms of disorders of consciousness. What the attribute "conscious" cannot do for the ethicist, the idea of a self and its remainders might help to achieve.

7.7 Conclusion

With the advent of the active para-clinical paradigms neuroscience suddenly found itself confronted with patients formerly thought to be completely unconscious, who now seemed to have some minimal consciousness preserved, but who were not *clearly*

[25]The Theory of Mind (ToM) discussed in cognitive psychology probably builds upon this more basic and general "theory" of a mind.

conscious either. This was a huge progress on the one hand, because it was now at least theoretically possible to find the 40% of minimally conscious patients among those who—based on *clinical* examination—had wrongly been diagnosed with a UWS. On the other hand, however, there are two follow-up questions to be answered: How much consciousness—if at all— is there "inside" those non-responsive patients? What does this mean in ethical regard?

As we have seen, there are four central problems regarding disorders of consciousness. *First,* we cannot simply ask the patients in question because they are frequently unresponsive. Therefore, consciousness can be *covered.* The *second* problem is that neuroscientific para-clinical examinations have inherent *limitations* regarding the human mind. These limitations are not so much due to the methods as such; rather they are a consequence of the special nature of consciousness. *Third,* if we do not take account of these limitations, we may fall victim of several fallacies that in the end might harm patients. The *fourth* problem is that the question of the moral value of consciousness remains unsettled. Is consciousness what matters? If we answer in the affirmative, we still do not know which kinds and degrees of conscious experience matter how much.

On this background, it seems best to base our ethical judgments regarding patients with a disorder of consciousness on the idea of a self. One may object that the term "self" is as fuzzy as the term "consciousness". Admittedly, this is true. In contrast to consciousness, however, the idea of a person's self as a neuroethical landmark allows to take not only different types of conscious contents into account, but non-conscious—or pre-conscious—information and functions as well. This will equip neuroethics with a richer account of what is at stake for patients with DOC. At least as long as it cannot be excluded with certainty that a patient has some residual consciousness and might regain fuller consciousness, we should take into account unconscious operations of the brain. The reason is that those unconscious processes might influence the way she feels on a (purely) phenomenal level.

Consciousness is more than brain function. Self-consciousness is more than conscious perception alone. A person's self is even richer and more complex than islands of self-consciousness.

References

Bayne, Tim, Jakob Hohwy, and Adrian M. Owen. 2016. Are There Levels of Consciousness? *Trends in Cognitive Sciences* 20 (6): 405–413. https://doi.org/10.1016/j.tics.2016.03.009.

Bender, Andreas, Ralf J. Jox, Eva Grill, Andreas Straube, and Dorothée Lulé. 2015. Persistent Vegetative State and Minimally Conscious State: A Systematic Review and Meta-Analysis of Diagnostic Procedures. *Deutsches Arzteblatt International* 112 (14): 235–242. https://doi.org/10.3238/arztebl.2015.0235.

Block, Ned. 1995. On a Confusion about a Function of Consciousness. *Behavioral and Brain Sciences* 18 (02): 227–247. https://doi.org/10.1017/S0140525X00038188.

Bruno, Marie-Aurélie., Audrey Vanhaudenhuyse, Aurore Thibaut, Gustave Moonen, and Steven Laureys. 2011. From Unresponsive Wakefulness to Minimally Conscious PLUS and Functional

Locked-in Syndromes: Recent Advances in Our Understanding of Disorders of Consciousness. *Journal of Neurology* 258 (7): 1373–1384. https://doi.org/10.1007/s00415-011-6114-x.

Cabrera, Laura Y., and Judy Illes. 2018. Balancing Ethics and Care in Disorders of Consciousness. *The Lancet Neurology* 17 (2): 112–113. https://doi.org/10.1016/S1474-4422(17)30402-7.

Cavaliere, Carlo, Carol di Perri, Steven Laureys, and Andrea Soddu. 2016. Instrumental Assessment of Residual Consciousness in DOCs. In *Neurotechnology and Direct Brain Communication: New Insights and Responsibilities Concerning Speechless but Communicative Subjects*, ed. by Michele Farisco and Kathinka Evers, 70–84. Explorations in Cognitive Psychology Series. London; New York: Routledge.

Corbetta, Maurizio, and Gordon L. Shulman. 2002. Control of Goal-Directed and Stimulus-Driven Attention in the Brain. *Nature Reviews Neuroscience* 3 (3): 201–215. https://doi.org/10.1038/nrn755.

Damasio, Antonio R. 1999. *The Feeling of What Happens: Body and Emotion in the Making of Consciousness*. Harcourt Brace.

Davies, Will, and Neil Levy. 2016. Persistent Vegetative State, Akinetic Mutism, and Consciousness. In *Finding Consciousness: The Neuroscience, Ethics, and Law of Severe Brain Damage*, ed. by Walter Sinnott-Armstrong, 122–34. Oxford Series in Neuroscience, Law, and Philosophy. New York: Oxford University Press.

Dennett, Daniel. 1992. The Self as a Center of Narrative Gravity. In *Self and Consciousness: Multiple Perspectives*, ed. Frank S. Kessel, Pamela M. Cole, and Dale L. Johnson, 103–115. Hillsdale, N.J: L. Erlbaum.

Eickhoff, S.B., M. Dafotakis, C. Grefkes, T. Stöcker, N.J. Shah, A. Schnitzler, K. Zilles, and M. Siebler. 2008. FMRI Reveals Cognitive and Emotional Processing in a Long-Term Comatose Patient. *Experimental Neurology* 214 (2): 240–246. https://doi.org/10.1016/j.expneurol.2008.08.007.

Farisco, Michele, and Kathinka Evers. 2017. The Ethical Relevance of the Unconscious. *Philosophy, Ethics, and Humanities in Medicine : PEHM* 12 (December). https://doi.org/10.1186/s13010-017-0053-9.

Fazekas, Peter, and Morten Overgaard. 2018. A Multi-Factor Account of Degrees of Awareness. *Cognitive Science* 42: 1833–1859. https://doi.org/10.1111/cogs.12478.

Fernández-Espejo, Davinia, and Adrian M. Owen. 2013. Detecting Awareness after Severe Brain Injury. *Nature Reviews. Neuroscience* 14 (11): 801–809. https://doi.org/10.1038/nrn3608.

Fins, Joseph J. 2015. *Rights Come to Mind. Brain Injury, Ethics, and the Struggle for Consciousness*. 1st ed. Cambridge University Press.

Fins, Joseph J., and James L. Bernat. 2018. Ethical, Palliative, and Policy Considerations in Disorders of Consciousness. *Archives of Physical Medicine and Rehabilitation* 99 (9): 1927–1931. https://doi.org/10.1016/j.apmr.2018.07.003.

Giacino, Joseph T., S. Ashwal, N. Childs, R. Cranford, B. Jennett, D.I. Katz, J.P. Kelly, et al. 2002. The Minimally Conscious State: Definition and Diagnostic Criteria. *Neurology* 58 (3): 349–353.

Giacino, Joseph T., M.A. Kezmarsky, J. DeLuca, and K.D. Cicerone. 1991. Monitoring Rate of Recovery to Predict Outcome in Minimally Responsive Patients. *Archives of Physical Medicine and Rehabilitation* 72 (11): 897–901.

Glannon, Walter. 2016. The Value and Disvalue of Consciousness. *Cambridge Quarterly of Healthcare Ethics* 25 (4): 600–612. https://doi.org/10.1017/S0963180116000335.

Greenberg, D.L. 2007. Comment on "Detecting Awareness in the Vegetative State." *Science* 315 (5816): 1221–1221. https://doi.org/10.1126/science.1135284.

Jennett, B., and F. Plum. 1972. Persistent Vegetative State after Brain Damage. A Syndrome in Search of a Name. *Lancet* 1 (7753): 734–737.

Johnson, L. Syd M.. 2016. Inference and Inductive Risk in Disorders of Consciousness. *AJOB Neuroscience* 7 (1): 35–43. https://doi.org/10.1080/21507740.2016.1150908.

de Jong, B.M., A.T. Willemsen, and A.M. Paans. 1997. Regional Cerebral Blood Flow Changes Related to Affective Speech Presentation in Persistent Vegetative State. *Clinical Neurology and Neurosurgery* 99 (3): 213–216.

Jox, Ralf J., and Katja Kuehlmeyer. 2013. Introduction: Reconsidering Disorders of Consciousness in Light of Neuroscientific Evidence. *Neuroethics* 6 (1): 1–3. https://doi.org/10.1007/s12152-011-9138-0.

Kahane, Guy, and Julian Savulscu. 2009. Brain Damage and the Moral Significance of Consciousness. *Journal of Medicine & Philosophy* 34 (1): 6–26. https://doi.org/10.1093/jmp/jhn038.

Kant, Immanuel. 1998. *Critique of Pure Reason*. The Cambridge Edition of the Works of Immanuel Kant. Cambridge ; New York: Cambridge University Press.

Laureys, Steven, Gastone G. Celesia, Francois Cohadon, Jan Lavrijsen, José León-Carrión, Walter G. Sannita, Leon Sazbon, et al. 2010. Unresponsive Wakefulness Syndrome: A New Name for the Vegetative State or Apallic Syndrome. *BMC Medicine* 8 (November): 68. https://doi.org/10.1186/1741-7015-8-68.

Laureys, Steven, and Nicholas D. Schiff. 2012. Coma and Consciousness: Paradigms (Re)Framed by Neuroimaging. *NeuroImage* 61 (2): 478–491. https://doi.org/10.1016/j.neuroimage.2011.12.041.

Levy, Neil. 2014. The Value of Consciousness. *Journal of Consciousness Studies: Controversies in Science & the Humanities* 21 (1–2): 127–138.

Levy, Neil, and Julian Savulescu. 2009. Moral Significance of Phenomenal Consciousness. In *Coma Science: Clinical and Ethical Implications*, ed. by Nicholas D. Schiff and Adrian M. Owen Steven Laureys, 177: 361–70. Progress in Brain Research. New York: Elsevier. https://www.sciencedirect.com/science/article/pii/S0079612309177257.

Mäki-Marttunen, Verónica, Mariana Castro, Lisandro Olmos, Ramón Leiguarda, and Mirta Villarreal. 2016. Modulation of the Default-Mode Network and the Attentional Network by Self-Referential Processes in Patients with Disorder of Consciousness. *Neuropsychologia* 82 (February): 149–160. https://doi.org/10.1016/j.neuropsychologia.2016.01.022.

Menon, D.K., A.M. Owen, E.J. Williams, P.S. Minhas, C.M. Allen, S.J. Boniface, and J.D. Pickard. 1998. Cortical Processing in Persistent Vegetative State. Wolfson Brain Imaging Centre Team. *Lancet (London, England)* 352 (9123): 200.

Monti, Martin M. 2012. Cognition in the Vegetative State. *Annual Review of Clinical Psychology* 8 (1): 431–454. https://doi.org/10.1146/annurev-clinpsy-032511-143050.

Monti, Martin M., Audrey Vanhaudenhuyse, Martin R. Coleman, Melanie Boly, John D. Pickard, Luaba Tshibanda, Adrian M. Owen, and Steven Laureys. 2010. Willful Modulation of Brain Activity in Disorders of Consciousness. *New England Journal of Medicine* 362 (7): 579–589. https://doi.org/10.1056/NEJMoa0905370.

Morin, Alain. 2006. Levels of Consciousness and Self-Awareness: A Comparison and Integration of Various Neurocognitive Views. *Consciousness and Cognition* 15 (2): 358–371. https://doi.org/10.1016/j.concog.2005.09.006.

Nachev, P., and M. Husain. 2007. Comment on "Detecting Awareness in the Vegetative State." *Science* 315 (5816): 1221–1221. https://doi.org/10.1126/science.1135096.

Nagel, Thomas. 1974. What Is It Like to Be a Bat? *the Philosophical Review* 83 (4): 435–450. https://doi.org/10.2307/2183914.

Naro, Antonino, Antonino Leo, Alfredo Manuli, Antonino Cannavò, Alessia Bramanti, Placido Bramanti, and Rocco Salvatore Calabrò. 2017. How Far Can We Go in Chronic Disorders of Consciousness Differential Diagnosis? The Use of Neuromodulation in Detecting Internal and External Awareness. *Neuroscience* 349 (May): 165–173. https://doi.org/10.1016/j.neuroscience.2017.02.053.

Neisser, Ulric. 1988. Five Kinds of Self-Knowledge. *Philosophical Psychology* 1 (1): 35–59.

Northoff, Georg. 2016. Doaes Task-Evoked Activity Entail Consciousness in Vetetative State? In *Neurotechnology and Direct Brain Communication: New Insights and Responsibilities Concerning Speechless but Communicative Subjects*, ed. by Michele Farisco and Kathinka Evers, 104–16. London; New York: Routledge.

Owen, Adrian M., Martin R. Coleman, Melanie Boly, Matthew H. Davis, Steven Laureys, and John D. Pickard. 2006. Detecting Awareness in the Vegetative State. *Science* 313 (5792): 1402–1402. https://doi.org/10.1126/science.1130197.

Pistoia, Francesca, Simona Sacco, Janet Stewart, Marco Sarà, and Antonio Carolei. 2016. Disorders of Consciousness: Painless or Painful Conditions?-Evidence from Neuroimaging Studies. *Brain Sciences* 6 (4). https://doi.org/10.3390/brainsci6040047.

Płonka, Beata. 2015. Neurobiology of Consciousness: Current Research and Perspectives. *Studia Humana* 4 (December): 23–38. https://doi.org/10.1515/sh-2015-0023.

Posner, Jerome B., Clifford B. Saper, Nicholas Schiff, and Fred Plum. 1983. *Plum and Posner's Diagnosis of Stupor and Coma*. Fourth Edition, 2007. Contemporary Neurology Series. Oxford, New York: Oxford University Press.

Raichle, M.E., A.M. MacLeod, A.Z. Snyder, W.J. Powers, D.A. Gusnard, and G.L. Shulman. 2001. A Default Mode of Brain Function. *Proceedings of the National Academy of Sciences* 98 (2): 676–682. https://doi.org/10.1073/pnas.98.2.676.

Schaefer, Michael, and Georg Northoff. 2017. Who Am I: The Conscious and the Unconscious Self. *Frontiers in Human Neuroscience* 11: 126. https://doi.org/10.3389/fnhum.2017.00126.

Schechtman, Marya. 2011. The Narrative Self. In *The Oxford Handbook of the Self*, ed. by Shaun Gallagher. Oxford ; New York, NY: Oxford University Press.

Schnakers, Caroline, Audrey Vanhaudenhuyse, Joseph Giacino, Manfredi Ventura, Melanie Boly, Steve Majerus, Gustave Moonen, and Steven Laureys. 2009. Diagnostic Accuracy of the Vegetative and Minimally Conscious State: Clinical Consensus versus Standardized Neurobehavioral Assessment. *BMC Neurology* 9 (1). https://doi.org/10.1186/1471-2377-9-35.

Shepherd, Joshua. 2018. *Consciousness and Moral Status*. Routledge Focus on Philosophy. New York: Routledge.

Staffen, W., M. Kronbichler, M. Aichhorn, A. Mair, and G. Ladurner. 2006. Selective Brain Activity in Response to One's Own Name in the Persistent Vegetative State. *Journal of Neurology, Neurosurgery, and Psychiatry* 77 (12): 1383–1384. https://doi.org/10.1136/jnnp.2006.095166.

The Multi-Society Task Force on PVS. 1994. Medical Aspects of the Persistent Vegetative State (1). *New England Journal of Medicine* 330 (21): 1499–1508. https://doi.org/10.1056/NEJM19 9405263302107.

Vanhaudenhuyse, Audrey, Vanessa Charland-Verville, Aurore Thibaut, Camille Chatelle, Jean-Flory L. Tshibanda, Audrey Maudoux, Marie-Elisabeth. Faymonville, Steven Laureys, and Olivia Gosseries. 2018. Conscious While Being Considered in an Unresponsive Wakefulness Syndrome for 20 Years. *Frontiers in Neurology* 9: 671. https://doi.org/10.3389/fneur.2018.00671.

Vanhaudenhuyse, Audrey, Athena Demertzi, Manuel Schabus, Quentin Noirhomme, Serge Bredart, Melanie Boly, Christophe Phillips, et al. 2011. Two Distinct Neuronal Networks Mediate the Awareness of Environment and of Self. *Journal of Cognitive Neuroscience* 23 (3): 570–578. https://doi.org/10.1162/jocn.2010.21488.

Vogeley, Kai, and Shaun Gallagher. 2011. Self in the Brain. In *The Oxford Handbook of the Self*, ed. by Shaun Gallagher, 111–36. Oxford ; New York, NY: Oxford University Press.

Wilkinson, Dominic, Guy Kahane, and Julian Savulescu. 2008. "Neglected Personhood" and Neglected Questions: Remarks on the Moral Significance of Consciousness. *The American Journal of Bioethics* 8 (9): 31–33. https://doi.org/10.1080/15265160802412486.

Marco Stier is "Privatdozent" (roughly equaling Associate Professor) and works as a scientific assistant at the Institute for Medical Ethics, History and Philosophy of Medicine at Münster University, Germany. His main fields of research are bioethics in general, neuroethics, philosophy of action, and philosophy of mind. Among his publications are two monographs on free will (*Verantwortung und Strafe ohne Freiheit*, mentis 2011; *Willensfreiheit: Bestimmt mein Gehirn oder bestimme ich?*, Reinhardt 2014) as well as several articles, e.g., "Normative Preconditions for the Assessment of Mental Disorder," in: *Frontiers in Theoretical and Philosophical Psychology* 2013/9, and "Potentiality in Bioethics," in: K. Engelhard, M. Quante (ed., 2018): *Handbook of Potentiality*).

Part II
From Medical Ethics to Theories of the Self and Autonomy

Chapter 8
Ethical Issues Concerning Patient Autonomy in Clinical Practice

Alfred Simon

Abstract The paper takes up the perspective of clinical practice and discusses the importance of respect for patients' autonomy from this angle. Until a few decades ago, physicians and nurses involved in ethical decision-making primarily considered the principles of beneficence and non-maleficence. Respect for the patient's autonomy clearly played a subordinate role. In traditional ethical codes such as the Hippocratic Oath or the Declaration of Geneva of the World Medical Association of 1948, no reference is made to the patient's wishes. This has changed radically: In line with the shift in values towards greater individuality and personal responsibility observed in the Western world in the second half of the twentieth century, patient self-determination became more and more important. Legal decisions and modern medical ethics emphasize that patient autonomy overrides what physicians and nurses consider best for the well-being of the patient. Informed consent is now a widely acknowledged normative standard in medical ethics, and patient autonomy finds its concrete expression in clinical practice in this concept of informed consent. The paper introduces the concept and discusses central issues concerning patient autonomy in clinical practice: What can health care professionals do to encourage patient autonomy? What are possible limits to patient autonomy? How can a patient's capacity to consent be assessed? How should health care professionals deal with patients who have temporarily or completely lost their capacity to consent? What is the importance of advance directives? What are possible problems concerning advance directives in clinical practice, and how can they be solved?

This chapter is a development of Simon and Nauck (2013).

A. Simon (✉)
University of Göttingen, Göttingen, Germany
e-mail: asimon1@gwdg.de

M. Kühler and V. L. Mitrović (eds.), *Theories of the Self and Autonomy in Medical Ethics*, The International Library of Bioethics 83,
https://doi.org/10.1007/978-3-030-56703-3_8

8.1 Introduction

For many centuries, medical and nursing activities were based primarily on the principle of beneficence and non-maleficence. Respect of the patient's autonomy played a minor role in comparison. The traditional codices such as the "Hippocratic Oath" or the "Declaration of Geneva" of the World Medical Association for 1948 do not contain a single reference to the will of the patient.[1] Well into the twentieth century, the relationship between the patient and the physician was marked by the paternalistic notion that not the patient, but the physician should, on the basis of his medical knowledge, decide what is best for the patient. The "merciful lie", i.e. the withholding of a bad diagnosis or prognosis with the well-meaning intention of "not dashing the hope" of the patient was considered ethically acceptable and at times required. Informed consent consultation was not meant to facilitate patient self-determination, but rather to convince the patient of the need for the suggested therapy, in order to get him involved, thus ensuring the success of the therapy (Katz 1984; Faden and Beauchamp 1986).

The picture has changed fundamentally in recent decades: As society has changed to increasingly value more individuality and self-responsibility, the patient's right to self-determination has become ever more important. It seems obvious nowadays, that patients not only are informed about their illness, but are actively involved in therapy decisions. Judicial rulings and official positions and statements by the medical profession emphasize the priority of the patient's will over what physicians and nurses regard as the well-being of the patient. Legal instruments like advance directives allow the (future) patient to exercise his right to self-determination in situations in which he himself is no longer capable of communicating and deciding.

The growing importance of the right to self-determination is creating new challenges for physicians and caregivers: They need to learn to properly listen to the patient and to act in the patient's best interest (which includes respecting his autonomy), rather than maximising diagnostic and therapeutic value. This can lead to difficult situations and ethical conflict in practice, e.g. the patient refusing a medically indicated treatment, or when physicians, caregivers, patient representatives and relatives disagree on how certain passages in the advance directive relate to further treatment. To resolve these conflicts, a careful balance needs to be struck between the principles of respecting autonomy, beneficence and non-maleficence[2] in the specific situation. A high level of ethical and communicative competence is required as well.

---

[1] Respect for patient autonomy was only introduced in the fourth revised version of the Geneva Declaration in 2017.

[2] Beauchamp and Childress (2013) list four ethical principles in their standard work *The Principles of Biomedical Ethics*: respect for autonomy, beneficence, non-maleficence and justice. This article focuses on the relationship between the first three principles, whereat the principles of beneficence and non-maleficence are summarised under the principle of care.

8.2 Autonomy and Self-determination

Autonomy is a key term in modern medical ethics. Like many terms in ethics, it can be interpreted in different ways (Feinberg 1986). As a descriptive term, it denotes the capacity for moral self-determination, i.e. the capability to decide and act according to one's own values and convictions. This capability deserves respect as an essential human feature, even if the actual ability for self-determination can in certain situations be limited or even lost. This touches on a second, normative meaning of autonomy, whereby every person is entitled to be recognised as a being capable of self-determination. This respect, which we owe our fellow human beings, prohibits hindering or ignoring self-determined decisions and actions, while also requiring these decisions to be supported, e.g. by appropriate informed consent procedures.

The literature mentions different conditions that must be met in order for a decision to be seen as self-determined and to be respected as such. Three continually accepted conditions are intentionality, understanding and noncontrol (Faden and Beauchamp 1986). Intentions are planned actions, regardless of whether the actor takes them for their own sake, as a means to an end or is merely accepting the consequences. This requires the person to understand and be able to deal with the significance of their actions, their consequences, opportunities and risks. Understanding, in turn, requires a sufficient level of insight and knowledge of the acting person. Mental illness, but also external influences such as manipulation or coercion, can constitute controlling influence.

It is disputed to what extent self-determined decisions must be authentic. A person acts authentically when, in critical reflection, they agree with the motives of their actions (Dworkin 1970; Frankfurt 1971). The question of authenticity arises in medicine, when e.g. the physician is confronted with patient decisions that considerably differ in terms of the subjective standards of values or compared to previous decisions of the patient. Such incoherent decisions can be an indication for limited or non-existent self-determination. A person can change their values and attitudes, however. Adhering too strictly to demands for authenticity in the above mentioned sense can result in wrongly dismissing unconventional decisions as not self-determined. One way to remedy this could be to discuss a seemingly incoherent decision with the patient, letting them critically reflect it, and subsequently accepting the result as self-determined regardless of the outcome.

8.3 Autonomy in Clinical Practice: Informed Consent

In clinical practice, autonomy of the patient manifests particularly in the concept of informed consent. Initially, it resulted as a legal demand for Medical Research Involving Human Subjects—i.e. in the "Nuremberg Code" from 1947—and has established itself in the entire medical field during the second half of the twentieth century (Faden and Beauchamp 1986). Medical ethics picked it up and promoted the demand for informed consent to complement and revise the traditional medical and nursing ethos, which had before been based primarily on the patient's well-being as defined by doctors and nurses.

Informed consent means that medical procedures, apart from emergencies, are only allowed with the consent of the informed patient or their proxy.

The aim of informing the patient is to put them into a position to make a self-determined decision for or against a treatment. This exceeds mere passing on of information, the decision-making ability of the patient is to be improved. The doctor[3] must therefore ascertain the understanding of the patient during an informed consent consultation, e.g. by asking the patient specific questions about a planned procedure. Furthermore, the patient needs to have the opportunity to ask questions himself, even to be encouraged to do so by the physician. In terms of content, the patient should be informed about anything that could be of importance for their decision for or against a treatment. This especially includes the diagnosis, the course and risks of the treatment, possible treatment alternatives as well as the consequences of not undergoing treatment. The informed consent procedure should occur in an individualized consultation tailored to the cognitive capabilities of the patient. Medical terms should be avoided when possible and otherwise explained. Ethical conflicts in the context of informed consent can arise when a patient refuses to acknowledge or ignores his situation and requests treatment which is not medically indicated, or rejects treatment even though it could avert harm from the patient. It may also occur that relatives demand the attending physician not to inform the patient of a malignant or not curable illness, because the patient ostensibly cannot handle the truth. In this case it is necessary to emphasize the right of the patient to be informed about their condition, to ensure their remaining life can be spent in maximum autonomy. The informed consent consultation is not a single talk with the patient but rather a process that requires a patient-physician relationship. Time, trust, professional, ethical, communicational, social and emotional competence are important factors for an informed consent consultation, as are the attitude and conduct of physicians and other professionally involved personnel. In the process, the physician takes on responsibility from the patient. The clear situation of being the respondent to the request for help of another person makes it possible to avoid the snares of an overbearing paternalistic, a purely autonomy based and also a service provider and customer based understanding, in favour of a relationship model.

A further substantial prerequisite of informed consent is the patent's capacity to give informed consent. This entails the capacity to understand the nature, significance and scope of the treatment on the basis of the informed consent consultation and to act on this basis, specifically to accept or reject suggested measures. The problem in determining the capacity to give informed consent in practice arises in that determining it is a yes/no decision, while the preconditions for it manifest themselves in a gradual fashion in the patient. A decision has to be made about if a threshold has been reached and different physicians can arrive at differing conclusions. It is also possible

[3]The following comments on informed consent apply not only to physicians, but also to representatives of other health care professions.

that a patient whose cognitive faculty is impaired can nevertheless comprehend and thus give informed consent to simple medical procedures, while not being able to understand and thus give informed consent to complex procedures. The capacity to give informed consent accordingly must exist or rather be ascertained for a specific procedure.

To determine the capacity to give informed consent, various assessment instruments, such as the MacArthur Treatment Competence Test (MacCAT-CR), have been developed. The MacCAT-CR tests the capacity to give informed consent based on the criteria of understanding of information, inferential thinking, ability to make and communicate decisions, and insight into the illness. Studies with the MacCAT-CR have shown a large proportion of people not capable of giving informed consent—both in psychiatric and internistic patients and healthy people in general (Grisso and Appelbaum 1998). This leads to the conclusion that the criteria to determine the capacity to give informed consent used in the MacCAT-CR are too strict, so that many patients classified as not having the capacity to give informed consent via the MacCAT-CR in fact might have been capable of giving informed consent, which goes against the original medical ethical intent of protecting patient autonomy. It needs to be furthermore noted from a medical ethics point of view that the assessment of a lack of the capacity to give informed consent is not merely an empirical result on the basis of a statistic procedure, but always a value judgement as well (Vollmann 2008).

As a result, one may say that tests like the MacCAT-CR can be helpful instruments to assess a patient's capacity to give informed consent especially in borderline cases, but they cannot replace the physician's judgement and they are not designed for everyday clinical practice. The author of the article rather suggests, while considering what has been said so far, the following questions to be used not as a test but as a guide to aid physicians in assessing the patient's capacity to give informed consent during the normal joint decision-making process (Simon and Nauck 2013):

- Did the patient understand the information conveyed during the informed consent consultation? Can they reproduce the important information in their own words?
- Does the patient have insight into the nature of their situation and illness? Do they understand they are sick and require treatment? Do they realize the gravity of their illness and the extent of their need for treatment?
- Can the patient use the information they understood to make an appropriate, sensible and reality-based decision?
- Can the patient communicate their decision to the physician?
- Is the patient acting authentically, specifically in concordance with their values, goals and attitudes? If not: Is the patient aware of the dissonance? Can they plausibly explain their change in attitude?

If the physician has to answer "no" to any of these questions, doubts as to the patient's capacity to give informed consent would appear to be justified, and further clarification should be made, e.g. in the context of a psychiatric consultation.

8.4 The Patient Incapable of Informed Consent

The right to self-determination fundamentally applies past the point of loss of the capacity to give informed consent. If the patient decided for or against a measure, while being capable of giving informed consent, it follows that the decision retains legitimacy also when the patient is not currently capable of decision-making. Examples of this include the consent of a patient to an operation under anaesthesia or the rejection of further dialysis by a patient with kidney disease. In either case, the loss of (current) self-determination due to anaesthesia or uremic coma does not alter the obligation to adhere to the patient's previous decision.

Advance directives, moreover, offer the opportunity to document desired treatment in the case of a loss of the capacity to give informed consent, or to designate a health care proxy. The first advance directives emerged in the late 1960s in the USA. They contained statements referring to medical measures at the end of life and were termed "living wills" (Kutner 1969). Their aim was to allow for a self-determined dying in the context of high-technology medicine. Advanced directives are not limited to measures at the end of life however, but can contain stipulations pertaining to the nature and extent of desired treatment in the case of dementia, permanent loss of consciousness or psychiatric illness (Simon 2013; Radenbach and Simon 2016).

In spite of advance directives being widely recognised, studies have repeatedly shown that living wills in particular do not fulfil the expectations placed in them, since they are not widespread, often not available when needed or not specific enough. On top of that, many physicians do not properly take them into account (Jox et al. 2015). The concept of Advance Care Planning (ACP) was developed against this background. This concept envisages that patients will be assisted in the preparation of advanced directives through qualified counselling services which reach out to them. These counselling services may be complemented by structural measures, such as the development of standardised forms, training of health care workers, or the development of procedural recommendations for dealing with advanced directives in a facility or on a facility overarching level across a specific region (In der Schmitten et al. 2016).

If the patient is not capable of giving informed consent and has not stated his treatment wishes in a living will, a health care proxy must decide for the patient. In many countries, legal regulation stipulates the patient themself, with help from a proxy directive (e.g. durable power of attorney for health care), can designate who can make decisions on their behalf in case of a loss of the capacity to give informed consent. A problem with decisions by health care proxies in practice is that the proxy does not always know the precise will of the patient. A study, in which patients and their proxies were asked about a fictitious treatment scenario, showed discrepancies between patients and their proxies in close to a third of decisions (Shalowitz et al. 2006). Studies on communication about desired treatment between patients and their health care proxies found evidence that conversations on this topic had not been adequate or non-existent and terms such as "vegetating" were interpreted quite differently (Clements 2009; Porensky and Carpenter 2008). It therefore seems sensible to include (possible) health care proxies in the process of advance care planning.

8.5 Autonomy in Emergencies

Medical decisions in emergency situations constitute a particular challenge. Since the patient is oftentimes not capable of decision-making and the urgency of emergency care does generally allow for neither the time and applicability of an existing advance directive to be checked, nor consent of a health care proxy to be attained, the physician is obligated to take the medically prescribed action in the assumption the patient will profit from and would consent to the measures. Should it later transpire that the action taken does not benefit the patient, e.g. because it only prolongs the dying process, or that it contradicts the patient's desire stated in the advance directive, the measures must be ceased. While it may be difficult and burdensome for many physicians to stop a measure they initiated and let the patient die, there is no ethical difference between not initiating and stopping a life-prolonging intervention: If the medical indication is missing or if the patient refuses to give their consent, both the initiation and the continuation of a measure are inadmissible. Should the attending physician be unable, for personal or ethical reasons, to comply with a change in therapy objective to cessation of therapy, e.g. ceasing invasive measures, they are not obligated to further treat the patient beyond an emergency situation. However, they must refer the patient to another physician who will continue treatment according to the patient's wishes (Wicclair 2011).

An emergency situation in the narrow sense of the term is to be distinguished from predictable life-threatening complications in the context of an existing disease. When they become foreseeable, the specific therapeutic measures which shall and shall not be taken when complications occur need to be discussed with the patient or—if they are not capable of decision-making—their health care proxy (Nauck and Alt-Epping 2008). The corresponding arrangements should be documented in the patient's records for the purpose of later verification and regularly checked. Here the usage of POLST-forms (Provider Orders for Life-Sustaining Treatment) can be helpful. These are standardized forms, in which the agreements made between the physician and the patient are documented in the form of a medical order in case of an emergency (Meier and Beresford 2009). This aims to ensure that the patient's treatment wishes are taken into account also when the emergency care is provided by another physician, e.g. the emergency physician.

8.6 Autonomy and Medical Paternalism

From what has been said so far, it is clear that autonomy in clinical practice cannot be achieved without the principle of care, which obliges doctors and nurses to promote the well-being of the patient and to avert harm from them. Patients who are limited in their capacity for self-determination especially depend on caring support in order to exercise their right to self-determination. The relationship between autonomy and care therefore seems unproblematic, as long as caring action aims at strengthening, maintaining or restoring autonomy. The crucial question, however, is to what extent interventions in autonomous patient decisions are justifiable by citing the principle of care. This question is the subject of the paternalism debate in medical ethics (Childress 1982; Schöne-Seifert 2009).

Paternalism refers to measures designed to protect the (ostensible) well-being of individuals against their present wishes and preferences. The ethical debate distinguishes different forms of paternalistic actions: Thus, in regards to persons who are recipients of these actions, one can differentiate between a strong and a weak paternalism: If the persons are capable of deciding their matters independently (persons capable of giving informed consent), it is a case of strong paternalism, in all other cases (persons not capable of giving informed consent), it is a case of weak paternalism (Feinberg 1971). In addition, regarding the measures being considered, a distinction can be made between a soft and a hard paternalism: Hard paternalism designates measures such as legal prohibitions or physical coercion preventing someone from doing something considered harmful to themselves. Soft paternalism conversely designates measures that aim to indirectly influence the preferences and behaviour of persons, be it by informing about risks or by changing the underlying conditions (e.g. increasing expenses) (Patzig 1989).

Examples of weak paternalism are promises by which parents induce their child to undergo a necessary but so far refused dental treatment (soft form), or the compulsory treatment of a patient incapable of judgement (hard form). In both cases, the justification for paternalistic behavior is that the person (child or mentally ill) affected by this act is currently unable to recognize or act in their long-term interests. On the condition that the measures applied are appropriate, i.e. that the goal they aspire to cannot be achieved with milder measures, weak paternalism is regarded by most authors as relatively unproblematic morally (Feinberg 1971; Patzig 1989; Schöne-Seifert 2009).

In contrast, strong paternalism constitutes an infringement on the patient's autonomy and is thus significantly harder to justify (Patzig 1989; Schöne-Seifert 2009). While evaluating the appropriate interventions, it is also relevant if they only consider the well-being of the patient (pure paternalism), or also the well-being of other persons (impure paternalism) (Dworkin 1972). Examples of a mixed form are advertising bans for cigarettes or smoking bans, which aim to protect not only the health of the smoker, but also the health of non-smokers. According to classical liberalism, only mixed forms of strong paternalism can be justified. Other authors also consider other forms legitimate, if the reduction of risk or the promotion of patient well-being achieved thereby is large, while the extent of infringement of autonomy is relatively small (Patzig 1989). This applies especially to the mild form of strong paternalism. An example for this would be a health insurance incentive program that aims to motivate patients to change unhealthy lifestyles, i.e. low level of exercise.

8.7 The Physician-Patient-Relationship

The strong weighting of the principle of autonomy compared to the principle of care accompanies a shift in the understanding of the roles of patients, physicians and care givers, as well as their relationships among one another. Medical literature describes different ideal-typical models of the physician-patient-relationship (Emanuel and Emanuel 1992), which can be amalgamated into three core models (Simon and Nauck 2013).

The paternalistic model emphasises the asymmetry of the therapeutic relationship resulting from patient and physician meeting as seeking help and (potentially) offering help. In the paternalistic model the physician plays the active role as the medical expert, while the patient, dependant, infirm and passive, literally patient, endures and follows the physician's instructions. This highly asymmetrical form of a relationship tends to emerge when the patient is existentially threatened, e.g. in an acute emergency, like treatment after an accident, in the intensive care unit, urgent operations or severe infections. Furthermore, it conforms with the need of many patients for a caring, omniscient, paternal or maternal figure who can take on important and momentous decisions in difficult existential situations.

In the shared decision-making or deliberative model, the image of the passive patient is replaced by that of the patient as a partner. It results from the experience that for prevention, limitation or betterment of illness and suffering, not only the physician, but also the patient himself needs to be involved. The patient takes an active part as a participating expert in the treatment procedures and thereby also takes on responsibility for the result. The shared decision-making model does not deny the asymmetry between patient and physician, but nevertheless recognises the patient's knowledge of and their responsibility for their illness and makes use of both. Ideally, the autonomy of the patient, and the duty of care of the physician complement each other in coming to a shared decision. This partnership does not end with the process of informed consent, but extends over the entire duration of the therapeutic relationship including the determination of the therapy goal, the anamnesis, diagnosis and prognosis, the decision for a certain treatment option, all the way up to the implementation and evaluation of the therapy, whereby a change of therapy or therapy target may occur, if the therapy goal is not reached, or even before, e.g. in case of adverse side effects.

In the consumer model,[4] the patient assumes the role of the customer, who is courted with individual health services and improved hotel services. In the relationship between customer and healthcare provider, the original asymmetry of the doctor-patient relationship is abolished or negated, which is why the customer model can only be valid for a limited range of health services, such as travel medical advice and prevention, the elimination or alleviation of minor discomfort, or improving interventions without reference to disease.

These three models of the doctor-patient relationship do not exclude each other, but rather complement each other and derive their legitimacy from the respective situation. Thus, a pregnant woman first appears as a customer in search of a suitable maternity clinic; she compares various offers and chooses the clinic whose offer best suits her expectations and wishes. In the prenatal consultation and during birth, the customer model is replaced by the partnership model; the partners involved (women, midwife, doctor) each bring their own expertise to the situation and support each other in their respective tasks and responsibilities. Should complications arise during birth—e.g. if an emergency caesarean section becomes necessary—the birthing woman is pushed into a more passive role, in which she is dependent on the professional expertise of the team attending her.

8.8 Summary and Conclusions

In recent decades, there has been a change in the relationship between patient, physician and caregiver, which can be described as a transition from a more caring-patronizing to a more partnership-oriented relationship model. This change was accompanied by a greater emphasis on the patient's autonomy and self-determination. However, putting too much emphasis solely on autonomy carries the risk of overburdening the patient in the case of difficult decisions. Seriously ill, old and disabled patients in particular are dependent on the caring help of doctors and nurses, relatives and friends. Autonomy and care are thus understood not as opposing, but complementary principles. The question is therefore not whether the autonomy of the patient should be respected, but how this can be done in an appropriate way (Marckmann 2011). In search for an appropriate balance of autonomy and care in the various phases of the therapeutic relationship, reflection on the following questions may serve as a guide (Simon and Nauck 2013):

Definition of the treatment goal:

- Which therapy goal is the patient aiming for?
- Is this realistic from a medical and nursing perspective?

[4]Summarises the "informative model" and the "interpretative model" from Emanuel and Emanuel (1992).

Medical indication:

- Which medical and nursing measures are required to achieve this goal?
- How do the participating physicians and nurses assess the potential benefits and the risk of harm from these measures?
- What are possible alternatives? How are their benefits and risk of damage assessed?

Decisions of the patient capable of giving informed consent:

- What information does the patient need for their decision?
- Did they understand the relayed information? Can they use it appropriately for their decision?
- Is the patient aware of the meaning and scope of their decision?
- Should the patient be offered further discussion and information opportunities?

Decisions by health care proxies when the patient is incapable of giving informed consent:

- Does a valid advance directive exist? How specific are the statements pertaining to the current treatment situation?
- Does the patient have a health care proxy? Does the urgency of the situation allow for the time to wait for the decision of the health care proxy?
- Are there any other indications of the ostensible will of the patient? How explicit and situational are they?
- Is there a consensus about the will of the patient? If not: can it be reached by a new conversation with the health care proxy or is there a need for a legal clarification?

References

Beauchamp, Tom L., and James F. Childress. 2013. *Principles of Biomedical Ethics*, 7th ed. Oxford: Oxford University Press.

Childress, James F. 1982. *Who Should Decide? Paternalism in Health Care*. Oxford, New York: Oxford University Press.

Clements, John M. 2009. Patient Perceptions on the Use of Advance Directives and Life-Prolonging Technology. *American Journal of Hospice and Palliative Medicine* 26 (4): 270–276. https://doi.org/10.1177/1049909109331886.

Dworkin, Gerald. 1970. Acting Freely. *Noûs* 4 (4): 367–383. https://doi.org/10.2307/2214680.

Dworkin, Gerald. 1972. Paternalism. *The Monist* 56 (1): 64–84. https://doi.org/10.5840/monist197256119.

Emanuel, Ezekiel J., and Linda L. Emanuel. 1992. Four Models of the Physician-Patient Relationship. *Journal of the American Medical Association* 267 (16): 2221–2226. https://doi.org/10.1001/jama.1992.03480160079038.

Faden, Ruth R., and Tom L. Beauchamp. 1986. *A History and Theory of Informed Consent*. New York, Oxford: Oxford University Press.

Feinberg, Joel. 1971. Legal Paternalism. *Canadian Journal of Philosophy* 1 (1): 105–124. https://doi.org/10.1080/00455091.1971.10716012.
Feinberg, Joel. 1986. *Harm to Self*. Oxford: Oxford University Press.
Frankfurt, Harry G. 1971. Freedom of the Will and the Concept of a Person. *The Journal of Philosophy* 68 (1): 5–20. https://www.jstor.org/stable/2024717.
Grisso, Thomas, and Paul S. Appelbaum. 1998. *Assessing Competence to Consent to Treatment. A Guide for Physicians and Other Health Professionals*. New York, Oxford: Oxford University Press.
In der Schmitten, Jürgen, Friedemann Nauck, and Georg Marckmann. 2016. Advance Care Planning: A New Concept to Realise Effective Advance Directives. *Palliativmedizin* 17: 177–195. https://doi.org/10.1055/s-0042-110711.
Jox, Ralf J., Jürgen in der Schmitten, and Georg Marckmann. 2015. Ethische Grenzen und Defizite der Patientenverfügung. In *Advance Care Planning. Von der Patientenverfügung zur gesundheitlichen Vorausplanung*, ed. Michael Coors, Jürgen in der Schmitten, and Ralf J. Jox, 23–38. Stuttgart: Kohlhammer.
Katz, Jay. 1984. *The Silent World of Doctor and Patient*. New York: Free Press.
Kutner, Luis. 1969. Due Process of Euthanasia: The Living Aill, A Proposal. *Indiana Law Journal* 44 (4): 539–554. https://www.repository.law.indiana.edu/ilj/vol44/iss4/2.
Marckmann, Georg. 2011. Selbstbestimmung bei entscheidungsunfähigen Patienten aus medizinethischer Sicht. In *Autonomie und Stellvertretung in der Medizin. Entscheidungsfindung bei nichteinwilligungsfähigen Patienten*, ed. Christof Breitsameder, 17–33. Stuttgart: Kohlhammer.
Meier, Diane E, and Larry Beresford. 2009. POLST Offers Next Stage in Honoring Patient Preferences. *Journal of Palliative Medicine* 12 (4): 291–295. https://doi.org/10.1089/jpm.2009.9648.
Nauck, Friedmann, and Alt-Epping, Bernd. 2008. Crises in palliative care—a comprehensive approach. *The Lancet Oncology* 9 (11): 1086–1091.
Patzig, Günther. 1989. Gibt es eine Gesundheitspflicht? *Ethik in der Medizin* 1 (1): 3–12.
Porensky, Emily K., and Brian D. Carpenter. 2008. Knowledge and Perceptions in Advance Care Planning. *Journal of Aging and Health* 20 (1): 89–106. https://doi.org/10.1177/0898264307309963.
Radenbach, Katrin, and Alfred Simon. 2016. Advance Care Planning in der Psychiatrie. *Ethik in der Medizin* 28 (3): 183–196. https://doi.org/10.1007/s00481-016-0399-3.
Schöne-Seifert, Bettina. 2009. Paternalismus. Zu seiner ethischen Rechtfertigung in Medizin und Psychiatrie. In *Jahrbuch für Wissenschaft und Ethik*, ed. Ludger Honnefelder and Dieter Sturma, 107–127. Berlin: De Gruyter.
Shalowitz, David I., Elizabeth Garrett-Mayer, and David Wendler. 2006. The Accuracy of Surrogate Decision Makers. A Systematic Review. *Archives of Internal Medicine* 166 (5): 493–497. https://doi.org/10.1001/archinte.166.5.493.
Simon, Alfred. 2013. Historical Review on Advance Directives. In *Advance Directives: Ethical Issues from an International Perspective*, ed. Peter Lack, Nikola Biller-Andorno, and Susanne Brauer, 3–16. Dordrecht: Springer.
Simon, Alfred, and Friedemann Nauck. 2013. Patientenautonomie in der klinischen Praxis. In *Patientenautonomie. Theoretische Grundlagen – Praktische Anwendungen*, ed. Claudia Wiesemann and Alfred Simon, 167–179. Münster: Mentis.
Vollmann, Jochen. 2008. *Patientenselbstbestimmung und Selbstbestimmungsfähigkeit. Beiträge zur Klinischen Ethik*. Stuttgart: Kohlhammer.
Wicclair, Mark R. 2011. *Conscientious Objections in Health Care*. Cambridge: Cambridge University Press.
World Medical Association. 1948, [7]2017. Declaration of Geneva. Accessed 10 Dec 2018. https://www.wma.net/policies-post/wma-declaration-of-geneva.

Alfred Simon, Prof. Dr. Phil. is Director of the Academy for Ethics in Medicine at the University Medical Center Göttingen. He is a member of the Ethical Advisory Board of the German Medical Association, speaker of the Ethics Task Force of the German Association for Palliative Care, and chairman of the Clinical Ethics Committee of the University Medical Center Göttingen. His research interests include clinical ethics, clinical ethical counseling, patient autonomy, patients' rights, advance directives, end-of-life decision making, ethics in palliative care, and ethics in psychiatry. His most recent publications include (among others): "Vorausverfügungen in der Psychiatrie," in: Sturma, D., Heinrichs, B., Honnefelder, L. (eds.): *Jahrbuch für Wissenschaft und Ethik*, vol. 22, Berlin: De Gruyter, 2018, 205–219; Zilles, D., Koller, M., Methfessel, I., Trost, S., Simon, A.: "Behandlung gegen den natürlichen Willen am Beispiel der Elektrokonvulsionstherapie. Klinische, juristische und ethische Aspekte," in: *Nervenarzt* 89 (3), 2018, 311–318; "Freiwilliger Verzicht auf Nahrung und Flüssigkeit ("Sterbefasten"). Ein Ausweg am Lebensende?" in: *Wege zum Menschen* 69, 2017, 487–497; Radenbach, K., Simon, A. "Advance Care Planning in der Psychiatrie," in: *Ethik in der Medizin* 28, 2016, 183–196; Radenbach, K., Falkai, P., Weber-Reich, T., Simon, A.: "Joint crisis plans and psychiatric advance directives in German psychiatric practice," in: *Journal of Medical Ethics* 40, 2014, 343–345; "Historical review on advance directives," in: Lack, P., Biller-Andorno, N., Brauer, S. (eds.): *Advance Directives: ethical issues from an international perspective*, Dordrecht: Springer, 2013, 3–16.

Chapter 9
A Patient's Right to Refuse Medical Treatment Through the Implementation of Advance Directives

Hajrija Mujović

Abstract Serbia's legislation on patient rights has largely been formulated in response to contemporary understandings of human rights. But the way in which this legislation has been implemented reveals a misunderstanding concerning the degree to which the rights explicated in the Patient Rights Act (Official Gazette of the Republic of Serbia No. 45. 2013) are applicable to patients receiving palliative care. There are issues particularly pertinent to terminally ill patients, including euthanasia (Article 28) and advance directives (Article 16) which need to be discussed and revised. First, according to the provisions, the term 'euthanasia' does not differentiate between passive and active forms of euthanasia, which means that both passive and active euthanasia is forbidden under Serbian law. However, in medical practice, there are situations in which passive euthanasia, e.g. withholding life-saving treatment, is permitted without sanction. The legality of this common practice is called into question by the current formulation of the Patient Rights Act. In practice, this conflict is not visible given the fact that situations like this are interpreted as falling under the patient's broad right to refuse medical treatment. Second, regarding advance directives, I will argue that, according to Serbian law, a dying patient is entitled to basic rights and human dignity. This entails a right to prepare one's own advance directive, and to have it respected. Accordingly, provisions to allow for this should be incorporated into Serbian legislation.

Keywords Medicine · Patient rights · Advance directives · Serbian law and ethics

9.1 Introduction

Today, especially in highly developed countries, a paternalistic approach to medical treatment has been largely abandoned, and is considered to be a relic of the past (Radišić 2008). The main reason for this is that the patient is now considered to be the subject rather than an object of a treatment, and there thus is a corresponding

H. Mujović (✉)
Institute of Social Sciences, Centre for Legal Research, Belgrade, Serbia
e-mail: hmujovic@idn.org.rs

M. Kühler and V. L. Mitrović (eds.), *Theories of the Self and Autonomy in Medical Ethics*, The International Library of Bioethics 83,
https://doi.org/10.1007/978-3-030-56703-3_9

tendency to prioritize the patient's wishes and desires concerning her treatment. It is often emphasized in the literature that this is about the understanding of a human being as a self-determining agent, who can take responsibility for her own life and treatment (Mujovic Zornic 2010). In addition to the traditional relationship of trust, the patient's autonomy is central to the understanding of the relationship between the physician and the patient as involving two equal partners engaged in a partnership of mutual rights and obligations (Radišić and Mujović Zornić 2004). The principle of the patient autonomy not only improves communication with the physician and emphasizes the patient's active role in decision-making, but also provides protection from an overly-authoritative physicians.

Although respect for autonomy is regarded by many ethicists as a fundamental value in the ethics of health care (Beauchamp and Childress 2013),[1] it has never been fully recognized as a legally protected interest. Instead, in law, patient autonomy is indirectly protected through two other, legally protected interests. The first of these interests is the physical integrity of the person. There is a provision forbidding the bodily injury, unless consent is given by the person concerned. Another interest is the wellbeing of the human body, which is protected by a provision on professional competence, and by prohibiting those who do not have medical qualification from practicing medicine (McHale and Fox 2007). Although respecting the patient's wishes also normally protects the patient welfare, disagreements can arise the doctor's notion of what is in the patient's best interests conflicts with her autonomous desires (Kaličanin 1999). The requirement to obtain the patient's consent under these circumstances serves to diminish the power of healthcare professionals who treat the patient. It is particularly important that the Medical law protects this right (Deutsch and Spickhoff 2014), given that health care professionals, above all the physician, are in a position of power over the patient, and that the patient is often in a vulnerable position, perhaps injured, subject to examination, or undergoing surgery. Given these power dynamics, requirements to provide information to the patient, and to obtain their consent, is key to maintaining an adequate balance of power (Mujović Zornić 2011).

9.1.1 *The Patient's Consent*

Under the Serbian legal system, in accordance with the law (Patient Rights Act 2013), each individual is guaranteed the equal right to health care, which should be provided in the best interest of the patient and with respect to his or her personal attitudes.[2] The patient has the right to freely decide on everything related to his or her life and health, except in cases when this directly threatens the life and health of others. Without the patient's consent, as a rule, no medical measures should be

[1] See more: the basic principles of biomedical ethics, based on moral judgment or moral belief, are respect for autonomy, nonmaleficence, beneficence and justice in dealing with the patient.

[2] Article 3, Patient Rights Act of RS.

taken. The concept of informed consent exists as a unified principle that includes a variety of legal issues in this area. It represents an ethical and legal expression of the human right of each person to have their autonomy or self-determination respected (Bjegović Mikanović et al. 2015). Under this framework, the requirement of consent effectively replaces trust as a cornerstone of the relationship between the patient and the health care professional. There is a legal presumption that patients are capable and competent to make their own decisions until the court decides otherwise.

A medical measure contrary to the will of the patient can only be taken in exceptional cases, which are established by law and which are in accordance with common standards of medical ethics, such as when the patient is a child, or has been deemed incompetent to make his own decisions.[3] The patient who is deemed capable of making his own decisions has the right to refuse any proposed medical measure, even if this measure is necessary to save his life. The healthcare professional is obliged to inform the patient of the consequences of his decision to reject the proposed medical measure and to obtain a written statement from the patient, indicating that he has been informed, and has understood this information. If the patient refuses to provide this written statement, an official note will be made. In either case, the health care professional must document the patient's (or the patient's legal representative's) consent to the proposed medical measure, as well as the refusal of that measure.[4] The Patient Rights Act abandons an older formulation of the patient's right to consent, which could have been more appropriately named 'self-determination'. The past provision of the Health Care Act included the patient's right to self-determination and free consent, which is omitted as the title of the relevant article in the new legislation, although the same text was retained.[5] The use of the term consent has been subject to criticism due to the fact that consent and self-determination are not exactly identical concepts, since self-determination can sometimes mean for a patient a plain refusal, i.e. reluctance and his extremely passive attitude when he does not want treatment or other measure (Mujović Zornić 2015). The difference is that the term consent does not contain refusal, whereas the term self-determination implies both making a decision, as well as refusing to make a decision. This is in fact the essence of the principle of informed consent and it is best understood and realized if the term self-determination is used. These are two related rights and reflect the autonomy of a patient.

This issue was also dealt with by the European Court of Human Rights in *Pretty v. United Kingdom*.[6] The European Court stated that the autonomy of the individual is an important principle which was interpreted to imply the right of the individual to conduct his life in accordance with his own decisions, including decisions to undertake activities that could be regarded as physically or morally harmful or dangerous to that individual (Pretty v. United Kingdom 2002). The court considered a situation in which a person refused medical treatment, and concluded that the imposition of

[3] Article 15, ibid.

[4] Article 17, ibid.

[5] Article 31, Health Care Act RS.

[6] App. No. 2346/2002.

medical treatment on an adult with adequate mental capacity, even if the treatment were necessary for the preservation of the patient's life, would have constituted a violation of the physical integrity of this person, guaranteed by Article 8 of the Convention. This case dealt with a woman with a severe and irreversible disease of the nervous system. The disease caused a rapid weakening of the muscles, a paralysis of her body, and significant interference with breathing and eating unassisted. Her mental and cognitive capacity was completely preserved. In the final stage of development of the disease, in order to end her extreme mental suffering and feelings of humiliation, she expressed a desire decide for herself the terms upon which she would die. Although suicide under English law does not constitute a criminal offense, she could not carry out this act on her own due to her health condition, and it was necessary that others help her. The English court rejected a plea from her husband to be allowed to assist her to die. In revisiting this decision, the European court proceeded from the view that the autonomy of an individual is an important principle on the basis of which the guarantees contained in Article 8 of the European Convention on Human Rights are interpreted. The pain and humiliation suffered by the applicant as a result of the progression of the disease was increasing, and her autonomous decision to decide upon the manner of her death should be honored (European Convention on Human Rights 1950). The overarching principle of the European Convention, as determined by the European court, is to respect the freedom and dignity of the individual.

9.1.2 Rejecting Treatment as a Form of Passive Euthanasia

At the General Assembly in Madrid in 1987, the World Medical Association issued a proclamation on this question, declaring that euthanasia, i.e. the act of deliberately ending the life of a patient, even at the patient's own request or at the request of close relatives, is unethical (WMA Declaration on Euthanasia 1987). This does not prevent the physician from respecting the desire of a patient to allow the natural process of death to follow its course in the terminal phase of sickness (Kaličanin 1999). This proclamation is sometimes misunderstood as applying to cases of passive euthanasia, i.e. the withdrawal of life-sustaining treatment. Additionally, the Hippocratic Oath was revised in 2016, bringing in the elements of respect for patient's autonomy and expressed will (Hippocratic Oath 2017).

Some legal theorists believe that the practice of euthanasia should be accepted in Serbia, especially if we are talking about passive forms of euthanasia, which is now allowed in many countries (Klajn Tatić 2002). In fact it is already conducted in practice in Serbia today, although physicians generally avoid using the term passive euthanasia. Despite the reluctance to talk directly about passive euthanasia, health care professionals in Serbia routinely respect patients' refusal of artificial life support, and honors patient requests to withdraw life support in case of terminal illness, including refusal of a feeding tube or artificial respiration. In addition, Serbian physicians routinely respect refusals to undergo palliative care, and allow some terminal patients to take pain medication when this will have the side effect of shortening the patient's life. All of these measures can be termed passive euthanasia, that is, they can be seen as allowing death to take place, rather than actively causing it. Active forms of euthanasia, for example, administering a lethal drug or injection, are not widely accepted, and are not a routine part of Serbian medical practice (though their merits can be debated).

We need to change the conversation around euthanasia in Serbia. Euthanasia should not be regarded in a negative light, or as an easy way out. The conversation about these issues, and their legal treatment, should take into account the increased importance placed on dying with dignity, and the right of patients to have their end-of-life wishes respected in medical ethics (Deutsch 1984). The goal is to avoid turning the treatment into torture, and to avoid this, all options must be considered.

According to Serbian criminal law,euthanasia is defined as 'deprivation of life from compassion' (Criminal Code 2016).[7] It is generally understood rather broadly as including forms of assisted suicide. Furthermore, it has a negative connotation—the potential for pressure from relatives, or for the abuse of euthanasia, is frequently emphasized in practice. This broad interpretation of euthanasia, and the focus on negative outcomes, obscures a key ethical issue. Namely, euthanasia must be discussed in the context of medicine, because the question of administering euthanasia only arises with severely, terminally ill patients. These patients are under the care of health care institutions. Once they are admitted to these institutions, and are diagnosed with a serious illness, the question of their treatment arises. Dying people seek medical assistance, either with extending their life, or with making their death more dignified and under their control (even if it comes sooner). The most important thing is to view these patients as human beings, not simply as the object of medical treatment, and to respect their will. When patients are suffering from a terminal illness, medical intervention can only do so much. Once the circumscribed nature of the patient's decision-making in these circumstances is made clear, the necessity of respecting his wishes to the greatest extent possible becomes more even more important. Once we abandon the assumption that euthanasia is necessarily a bad thing, and we view it in light of the importance of respecting the patient's decisions in a medical context, euthanasia starts to become more ethically justifiable.

In addition, there is legal backing for prioritizing the will of the patient, when it comes to passive euthanasia. As discussed above, the Patient Right Act protects the patient from any unwanted medical treatment, in all but exceptional cases. If there is doubt concerning the patient's will (if he is, for example, in a coma), the appropriate course of action is to be determined by a suitable legal representative, in consultation with those closest to the patient, with the aim of establishing and executing what the patient would have wanted. Of course, determining what the patient would have wanted can be a difficult task. In these cases, Serbian law does not require complete certainty, but only evidence pointing towards what the patient would have wanted (even if they had never overtly expressed a desire concerning these specific circumstances). According to Article 3 of the Patient Rights Act, medical professionals

[7] Article 117, Criminal code of RS.

are required to uphold professional standards, which involve prioritising the best interests of the patient, while respecting his personal wishes. In this context, procedural law must be considered as well. Acting on uncertain estimations of a patient's wishes might prove difficult for some medical professionals. But given the legal and ethical importance of respecting the patient's wishes, establishing these wishes with complete certainty is too high a standard, that will lead in practice to the patient's will often being ignored or overridden.

9.1.3 Specific Aspects of the Serbian Law

Advance directives are a good way to avoid many of these aforementioned issues concerning uncertainty, and they should be respected, if they meet the legal requirements. Serbian health law, however, does not expressly regulate advance directives. There is no provision in the Patient Rights Act that focusing on or defining what is considered a legitimate advance directive. However, this act allows that the patient may make decisions about future medical treatment. The patient is also permitted to appoint a close relative or guardian to make decisions about treatment in case of future incapacity. Article 16 states that a "patient has the right to designate a person who will give consent on his behalf, or who will be informed about undertaking medical measures, in case the patient becomes incapable of taking the decision about giving consent" (Patient Right Act 2013), A subsequent paragraph of this article focuses on the circumstances under which the patient's will should be carried out, or overridden—it is clear, here, that this is referring to a statement of will which is given in advance. This article does not give sufficient legal guidance on advance directives, because it lacks important detail. It does not say, for example, which form advance directives must take. Although the existing provisions for advance directives represent a positive step, better ensuring that the autonomy of the patient is respected, these laws need to be further elaborated to address these issues and avoid ambiguity. Without additional detail, this provision may not function to adequately protect the patient's autonomy.

In current practice, many patients are not subjected to life extending treatment when there is no chance of recovery. Although allowing patients to die in this manner is clearly a type of passive euthanasia, it is never referred to as such by medical professionals in Serbia. Lack of available medical equipment can also be the cause of a failure to treat a dying patient, or health care professionals might also refuse to intervene due to fear of possible legal sanctions. In addition, it is often overlooked that a decision whether to treat a patient, or to refrain from treatment, should not just be made according to medical standards and the judgments of healthcare professionals, but need to be made in consultation with the patient or the patient's appointed guardian, as the law demands.

Euthanasia, in all its forms, is only permissible when it is voluntary. On the other hand, euthanasia requires the involvement and assistance of the physician. Euthanasia is not adequately regulated in Serbia, because the law that prohibits euthanasia does not differentiate between active and passive forms of euthanasia. However, in practice, passive euthanasia is routinely conducted, although it is not labelled as such. The reasons for the legal ban on passive euthanasia, and the medical reluctance to refer to it, might come down to standards in medical ethics, which consider these acts as killing (Kaličanin 1999). The formulation of the Serbian law, in its failure to distinguish between active and passive euthanasia, is not able to adequately deal with the many and complex forms of euthanasia.

We need a more fundamental consideration of the purpose and place of euthanasia, and of life extending treatment, in Serbia, and to devise uniform standards on the basis of this consideration, to ensure that patients across Serbia receive the same standard of care. In addition to the considerations about euthanasia which I have explored above, we should consider the purpose of life extending treatment. Is extending a patient's life in itself an aim of medicine? Given the increased abilities of life sustaining technologies to keep patients alive without a corresponding good quality of life, I contend we should look more to the standards of medical ethics—prioritizing the patient's wellbeing, dignity, and autonomous choice. These standards advocate the introduction of methods for choosing one's own manner of death.

9.2 Conclusion

The autonomy of the patient is undoubtedly a central value in medicine. Barring exceptional circumstances, a patient's consent is essential to the legitimacy of any medical intervention. Advance directives were designed to solve the problem of obtaining consent from patients that were no longer able to make decisions, further moving us towards the goal of respecting the patient's autonomous decisions in as many cases as possible, and away from medical paternalism.

Euthanasia, in both its passive and active forms, contravene Serbian criminal law. But civil law is less explicit about euthanasia, and this causes problems in medical practice, where the right to self-determination is paramount. In addition, passive euthanasia, which is technically included in the law contravening euthanasia, although this is rather ambiguous, is at the same time accepted as professional practice in some branches, for example in oncology or neonatology. There are no uniform standards for the treatment of dying patients, so patients can receive very different treatment, and be presented with very different options, depending on their physician and institution. In addition, palliative care is still a new and developing field.

The Serbian law is missing crucial detail, which leaves it unable to deal with the complexities of euthanasia, particularly the distinction between passive and active euthanasia. The law must be reformed to more adequately deal with the vast range of situations which involve patient's wish to end his's own life, in their variety and complexity.

The notion of self-determination also needs to be revisited in Serbian law—as it stands, there is not enough detail on what a right to self-determination entails for patients. The Serbian law contains a legal provision concerning progressive diseases, where it is taken for granted that the patient loses the capacity to decide for himself as the disease progresses. He is thus able to appoint a surrogate decision maker, or to write an advance directive. This provision is within the framework of the regulation of medical decisions in general, and does not refer to specific situations. It is within a general provision that refers to all patients, including, but not only referring to, patients with a terminal illness. More protection is needed for these terminal patients—we require a legal framework that is more specific concerning the circumstances and condition of the terminally ill patient. It is only then that health care professional will have adequate guidance in dealing with these situations. Here, as in the rest of medicine, the will of the agent should be key to the legitimacy of such procedures, even when these desires are expressed in advance of the development of the condition, in the form of an advance directive. In order to ensure that advance directives can be effectively brought into operation, these provisions should include a legal provision obligation health care professional to inform patients of their right to exercise their self-determination through an advance directive. Advance directives should become a routine part of a treatment plan.

Although there are some provisions for advance directives in Serbian law, these provisions need to be revisited and strengthened, because as it stands, many patients are not aware of the option of creating advance directives, and many physicians are confused about whether and how to apply them. As a result, they are rarely implemented. In including advance directives in a general provision of the Patient's Rights Act, rather than dedicating a specific provision to them, and in failing to mention them by name, the current Patient Rights Act fails to provide adequate, functional guidance on this matter. Such directives should be accorded greater prominence within the Serbian law, and should become a key part of medical practice, for the benefit of all.

References

Beauchamp, T.L., and J.F. Childress. 2013. *Principles of Biomedical Ethics*. New York: Oxford University Press.

Bjegović Mikanović, V., J. Santrić and J. Overal (eds.). 2015. Human Rights in Patient Care—Serbia, Chapters 6, 7. Belgrade: Medical School University of Belgrade. https://cop.health-rights.org/PractitionerGuides.

Criminal code. 2005. Official Gazette of the Republic of Serbia No. 85/2005, 88/2005—rev., 107/2005—amd., 72/2009, 111/2009, 121/2012, 104/2013, 108/2014 and 94/2016.

Deutsch, E. 1984. Euthanasia: A German View. In: *Euthanasia*, ed. A. Carmi, 92. Heidelberg: Medicolegal Library Springer Verlag.

Deutsch, E., and A. Spickhoff. 2014. *Medizinrecht - Arztrecht, Arzneimittelrecht, Medizinprodukterecht und Transfusionsrecht*. Berlin Heidelberg: Springer-Verlag.

European Convention on Human Rights. Rome 1950.

Health Care Act. 2005. Official Gazette of the Republic of Serbia No. 107/2005, 72/2009—Sec. Law, 88/2010, 99/2010, 57/2011, 119/2012, 45/2013—Sec. Law, 93/2014.

Hippocratic Oath. 2017. The WMA Adopted a Revised Version at its 68th General Assembly in Chicago Oct 13, 2017. http://hudoc.echr.coe.int/eng#{%22appno%22:[%222346/02%22],%22itemid%22:[%22001-60448%22]}.

Kaličanin, P. 1999. *Medicinska etika i medicinsko pravo*. Beograd: Institut za mentalno zdravlje.

Klajn Tatić, V. 2002. *Lekareva pomoć neizlečivo bolesnom pacijentu – etički i pravni problemi*. Beograd: Institut društvenih nauka.

Mc Hale, J., and M. Fox. 2007. *Health Care Law—Text and Materials*. London: Sweet & Maxwel.

Mujović Zornić, H. 2010. *Pojam i razvoj pacijentovih prava*. In: *Pacijentova prava u sistemu zdravstva*, edited by Jancic O, 13. Beograd: Institut društvenih nauka.

Mujović Zornić, H. 2011. Challenges of traditional bioethical principles in the implementation of contemporary standards of medical law. *Filozofija i društvo* 4:71–79.

Mujović Zornić, H. 2015. "Autonomija pacijenta – pravo na samoodređenje". *Pravni život* 9:277–292.

Patient Rights Act. 2013. Official Gazette of the Republic of Serbia No. 45.

Pretty v. United Kingdom. 2002. App. No. 2346.

Radisić, J. 2008. *Medicinsko pravo*. Beograd: Nomos.

Radisić, J., and H. Mujović Zornić. 2004. *Pomoć pacijentima u ostvarivanju njihovih prava: Zapadna Evropa kao uzor Srbiji*. Beograd: Akademija.

WMA Declaration on Euthanasia. 2005. Adopted by the 39th World Medical Assembly Madrid Spain October 1987 and reaffirmed by the 170th WMA Council Session Divonne-les-Bains France May 2005 and reaffirmed by the 200th WMA Council Session Oslo Norway April 2015.

Hajrija Mujović is Principal Research Fellow, lawyer with Ph.D. degree in Medical Law, and Head of the Centre for Legal Research at the Institute of Social Sciences in Belgrade. She is the project leader and adviser for many legal issues in health care, visiting professor in several Law Schools in Serbia and abroad within the Balkan region (Croatia, Bosnia), and a teacher in the master studies of Neuroethics at the School of Medicine, Belgrade University. She is also a member of the ethical board. Her specific areas of scientific research are: Medical Law, Patients' Rights, Medical liability, Pharmaceutical Law, and Public Health Law. Her publications iclude: Commentary on The Act on Prevention and Diagnosis of Genetic Diseases, Genetically Conditioned Anomalies and Rare Diseases, Belgrade, 2019, p.123, Medical expert witness' work in the theory and practice of Medical law, Belgrade, 2019, p. 310; "The Medical Law Aspects of transportation medicine's regulation," Teme, vol. XLII, 3/2018, https://doi.org/10.22190/TEME1803019M pp. 1019-1041; "Patient's Rights in Mental Health: Problems of Residential Care in Serbia," Medicine and Law, World Association for Medical Law, vol. 37, 2/2018, p. 15; Human Rights and Patient Care—Serbia (Chaps. 6 and 7), co-authorship with V. Bjegović Mikanović et al., ed. V. Bjegović Mikanović, J. Šantrić, J. Overal, Medical School, University of Belgrade, Belgrade, 2015, Chaps. 6 and 7, p. 413; "Personal rights in the field of Genetic diagnostics—New legislation of Serbia," in: Current issues of Civil and Commercial Law and Legal practice, Law School University of Mostar, ed. V. Kazazić, Bosnia and Herzegovina, Mostar, 2015, pp. 439–452; Human Rights and values in Biomedicine—Aspects of decision-making in Health care, (editor and co-author), Belgrade, 2014, p. 305; "Human Reproduction and Liability for 'new life' in the light of ethical and medico-legal consideration," in: Applying Ethics, Mediterranean Publishing, Hellenic Society for Ethics, Novi Sad, 2014, pp.132–153.

Chapter 10
Prolonged Autonomy? The Principle of Precedent Autonomy and the Binding Force of Advance Directives in Dementia

Oliver Hallich

Abstract Dementia has become a central problem of health care in modern societies. Currently, 47.5 million people worldwide have the disease, and it is projected that 75.6 million people will have it in 2030. Against this background, the question of the binding force of advance directives in dementia will be of growing importance in future years. Defenders of the view that dementia advance directives should be seen as binding often invoke the principle of precedent autonomy to support their view. According to this principle, it is the right of a competent individual to make decisions for a later time once competence has been lost. However, the paper criticizes this principle of precedent autonomy and argues that it cannot be upheld because there are some cases in which it is morally legitimate to act against the will of the competent patient that is expressed in an advance directive pertaining to a later life stage in which competence has been lost. First, the way for a criticism of the principle of precedent autonomy is paved by laying bare its fundamental premise, namely the "dominance thesis," according to which critical interests dominate experiential interests in cases of conflict between these two kinds of interests. In a second step, various ways in which critical and experiential interests may relate to each other are distinguished, and it is argued that the dominance thesis holds true only with regard to some but not all of these constellations. In focusing on a specific scenario with regard to which the dominance thesis turns out to be false, the paper argues that in some cases of a conflict between critical and experiential interests the latter ones negate the normative authority of the former. Based on a rebuttal of two objections to this view, it is concluded that we should answer the normative question of whether advance directives in dementia should be seen as binding with a qualified "no."

O. Hallich (✉)
University Duisburg-Essen, Duisburg, Germany
e-mail: oliver.hallich@uni-due.de

M. Kühler and V. L. Mitrović (eds.), *Theories of the Self and Autonomy in Medical Ethics*, The International Library of Bioethics 83,
https://doi.org/10.1007/978-3-030-56703-3_10

10.1 Introduction

Dementia has become a central problem for health care in modern societies. Currently, 47.5 million people worldwide have the disease, and it is projected that 75.6 million people will have it in 2030. The number of cases of dementia is estimated to more than triple by 2050.[1] Against this background, the question of the binding force of advance directives in dementia will be of growing importance in future years. Defenders of the view that dementia advance directives are morally binding often invoke the *principle of precedent autonomy* to support their view. According to this principle, it is the right of a competent individual to make decisions for a later time when competence has been lost. It is a specification of the more general principle of autonomy. Broadly speaking, the principle of autonomy claims that it is the right of the individual to make choices and take actions that reflect her own values and beliefs and that are free from controlling influences that determine her actions.[2] Extended over time and applied to cases of end-of-life decision-making, it requires respecting every individual's right to determine treatment in later stages of her life, including stages in which she has lost competence.[3]

In this article, I will cast a critical light on the principle of precedent autonomy. Though I will not dismiss it as wrong *sans phrase*, I contend that it does not have the justificatory role for medical decisions that many philosophers ascribe to it. In particular, I argue that it provides no justificatory basis for overriding the current interests of the demented patient if these run counter to the interests the patient formerly expressed in her advance directive. In part (1), I pave the way for a criticism of the principle of precedent autonomy by laying bare its fundamental premiss, which I shall call the "dominance thesis", according to which in cases of a conflict between critical and experiential interests, the former override the latter. In part (2), I distinguish between various ways in which critical and experiential interests may relate to each other, and I argue that the dominance thesis holds true only with regard to some, not all of these constellations. In part (3), I focus on a specific scenario in which the dominance thesis turns out to be false: in some cases of a conflict between critical and experiential interests, or so I argue, the latter ones negate the normative

[1]These figures draw on estimations of the World Health Organisation (2015). The Alzheimer's Association (2015) provides detailed U.S.-related data on the prevalence of dementia. For its prevalence in the U.K., see the most recent annual report of the Alzheimer's Society (2015). For a general overview on the prevalence of dementia, see Jennings (2015, 437).

[2]For a classical exposition of the principle of autonomy, see, e.g., Feinberg 1986, 27–97 (Chaps. 18, 19), Beauchamp and Faden (1986, 235–273) (Chap. 7), Beauchamp and Childress (2013, 101–149) (Chap. 4).

[3]The most prominent defenders of the principle of precedent autonomy are Rhoden (1990), Dworkin (1994, 218–241) (Chap. 8), Cantor (1992), Cantor (2005), Quante (2002, 268–295) (Chap. 7), Davis (2002), Davis (2004), Davis (2009), King (2007, 76–107) (Chap. 3). The most persistent critics of the principle of precedent autonomy are Rebecca Dresser and John Robertson; see Dresser (1984), Dresser (1986), Dresser (1994), Dresser (1995), Dresser and Robertson (1989), Robertson (1991); see also Buchanan and Brock (1990, 101–112). Instead of "precedent autonomy", some authors use terms like "prospective autonomy", "future-oriented autonomy" (Cantor 1992) or "prolonged autonomy" (Quante 2002, 268–295).

authority of the former ones. In part (4), I attempt to rebut three objections to this view. Part (5) contains some concluding remarks concerning the normative question of whether advance directives in dementia are morally binding or not—to which my answer is a qualified "No".

10.2 The Principle of Precedent Autonomy and the "Dominance-Thesis"

The principle of precedent autonomy rests on a premiss which I propose to call the "dominance thesis". The dominance thesis holds that in cases of a conflict between two kinds of interests, namely critical and experiential interests, interests of the first kind override or "dominate" those of the second kind. The distinction between critical and experiential interests was famously introduced by Ronald Dworkin.[4] Both kinds of interests, Dworkin argues, are "reasons people have for wanting their lives to go one way rather than another".[5] Experiential interests are interests whose fulfillment consists in experiences, be it the experience of listening to a Mozart-symphony, of enjoying food, of meditating over a complex chess problem or of having sex. These interests are time-sensitive in the sense that a person believes the experience to be desirable only when she actually has the experiential interest. Critical interests, by contrast, stem from judgements about what makes a life valuable. The fulfillment of these interests is not tied to actual experiences. Rather, they are interests that define our life plans and determine how we think we ought to live. They serve to guide and control our experiential interests. A person may have a critical interest in promoting science, in leading a healthy life, in being useful to her fellow beings, in maximizing her own welfare, in maintaining close friendships or in retaining her sexual purity. These interests are normative in the sense that they relate to projects the person thinks she ought to pursue and which she thinks make her life valuable and worthwhile.

It is possible that a person can lose her ability to form critical interests while still being able to entertain experiential interests. In some stages of dementia, for example, the demented patient can lose the ability to form critical interests, but may still have experiential interests. She may enjoy good food and feel basic pleasures, such as pleasure involved in attempting to communicate or in plays. Her life may be very similar to that of a normal five-year-old child. These experiential interests need not, but may run contrary to the critical interests the once competent person expressed in her advance directive. Whereas the competent person may have expressed the wish that, should she ever be in a state of dementia, life-sustaining treatment ought to be withheld, the experiential interests of the demented patient may be indicative of the wish that life-sustaining treatment be provided and that she be given the opportunity

[4]Dworkin (1994, 201–211). For a more detailed discussion of this influential distinction, see, e.g., Dresser (1995, 35–36), Jaworska (1999, 105–138).

[5]Dworkin (1994, 201).

to continue her life. In other words, the critical interests she once had as a competent person may clash with the experiential interests she now has as a demented patient.[6]

The dominance thesis claims that, should critical interests collide with experiential interests, the former interests trump the latter ones. Thus, we ought to respect them even at the price of acting against a person's present experiential interests. Evidently, this view has an impact on the question of the validity of advance directives pertaining to a state of dementia. Supposing that the dominance thesis is true, the experiential interests of the demented patient, should they be in conflict with former critical interests expressed in the advance directive, should be accorded less weight than the critical interests. Contravening experiential interests cannot then eliminate the normative authority of critical interests, and when treating the demented patient we ought to respect the preceding autonomous choices of the competent person as expressed in the advance directive. The requirement to respect her precedent autonomy would persist even though she is in a state in which she is no longer capable of making autonomous decisions. Any attempt to negate her critical interests on the basis of her experiential interests would, on this reading, be a violation of her autonomy.[7]

10.3 The Relation Between Critical and Experiential Interests—Various Constellations

Is thedominance thesis true? Should we therefore honor the critical interests embodied in advance directives in dementia? I do not think there is a clear-cut yes-or-no answer to these questions. Rather, I will adopt a context-dependent approach: as dementia is a complex phenomenon that comprises different stages,[8] we should allow for the possibility that the dominance thesis holds true in some cases but not in others.[9] It is therefore imperative to take a step back and to disentangle the various

[6]This clash does not necessarily imply that the demented patient is no longer identical with the competent person who issued the advance directive. In this paper, I do not enter the complicated debate on advance directives and the personal identity problem. For a detailed discussion of this problem, see, e.g., Buchanan (1988), Buchanan and Brock (1990, 152–189) (Chap. 3), DeGrazia (2005, 159–202) (Chap. 5).

[7]Most recently, Buller (2015, esp. 702), embraces this view.

[8]When talking about different stages of dementia, I refer to the "global deterioration scale for assessment of primary degenerative dementia" as developed by Reisberg et al. (1982). Reisberg et al. distinguish between the initial "forgetfulness phase", in which the deficit is primarily subjective but verifiable, the intermediate "confusional phase", in which the deficit is apparent to an objective observer, and the "late dementia phase", in which the patient can no longer survive without assistance. The first phase comprises stages (1) ("no cognitive decline") and (2) ("very mild cognitive decline"), the second stages (3) ("mild cognitive decline") and (4) ("moderate cognitive decline"), the third stages (5) ("moderately severe cognitive decline"), (6) ("severe cognitive decline") and (7) ("very severe cognitive decline") of the disease.

[9]For a defence of a context-dependent approach to end-of-life decisions, see also Menzel and Steinbock (2013, esp. 485), Menzel and Chandler-Cramer (2014, esp. 33–34).

ways in which critical and experiential interests may relate to each other. Broadly speaking, we can distinguish between "no-conflict-cases" and "conflict-cases".

(A) *No-conflict-cases*: There are some scenarios in which critical and experiential interests do not conflict at all. In these cases, the dominance thesis, which presupposes such a conflict, is not applicable. We can distinguish between three sub-scenarios:

(A1) *Concordance*: Critical interests may be consonant with experiential interests. A person may have a critical interest in leading a healthy life and have an experiential interest in eating salads and vegetables. Another person may aspire to leading an ascetic and "pure" life devoted to religious ideals, and this critical interest may coincide with the absence of any sexual desire and an experiential interest in abstaining from sexual relationships. In these cases, it would be inappropriate to talk of critical interests dominating experiential interests. Both kinds of interests match, and the question whether critical interests dominate experiential interests is inapposite.

(A2) *Post-mortem interests*: It may also be the case that there is no conflict between critical and experiential interests because a person's critical interests relate to a state in which she has no interests whatsoever. A person may have a critical interest in increasing the welfare of others and therefore consent to donate her organs to those in need of them after her death. Or she may, in her last will, declare her wish that all her property be given to charity organizations. These interests pertain to states in which the person who holds them will be dead, i.e. will have no interests of any kind. One is inclined to say that in these cases, we ought to respect the person's critical interests after her death. However, it would be wrong to say that her critical interests override her experiential interests, since there are no experiential interests that could be overridden. In the absence of a conflict between critical and experiential interests, the question whether we ought to subordinate experiential interests to critical ones does not arise.

Post-mortem directives of this kind provide an important insight: critical interests do not (necessarily) cease to be authoritative if these interests cease to exist. If there is an obligation to honor them, the mere fact that their fulfillment or non-fulfillment no longer affects the well-being of the subject who had these interests because these interests no longer exist does not cancel the obligation.[10] With regard to advance directives in dementia, this means that the mere fact that a late stage demented person, having lost competence, no longer has critical interests does not imply that there is no obligation to respect the critical interests of the once competent person. Rather, additional argument is needed to show that the preceding critical interests of

[10] For a defence of this claim, see also McMahan (2002, 497). Menzel and Steinbock write that critical interests still "exist […] even after one no longer experiences them" (2013, 489). Buchanan labels these interests "surviving interests" (1988, 287; see also Buchanan and Brock 1990, 162 164). According to Davis, interests can "survive" as long as it is nomologically possible for the interest holder to regain competence (2009, 367–371). To say that interests still "exist" after the death of the interest holder strikes me as counterintuitive. It would mean that we have no way of setting limits to the persistence of interests. Do we really want to say that Julius Caesar, anno 2015, still has an interest in not being murdered?

the once competent person ought not to be respected when this person is deprived of her cognitive abilities, i.e. loses her capacity to form critical interests.

(A3) *No contravening interests*: A person's critical interests may relate to a state in which the person, lacking the ability to form critical interests, still has experiential interests, but only experiential interests that do not run contrary to the preceding critical interests. Some advance directives in dementia exemplify this possibility. A person may write an advance directive in which she requests to withhold water and food by mouth in late stage dementia. In that state of advanced dementia, she may either resist eating and drinking—that is, her current interests may align with the directive—or she may not resist eating and drinking but not desire to eat and drink either. In this latter case, her will, based on her experiential interests, is neutral with regard to the will that she expressed in the advance directive. The person's current will does not align with her former critical interest, but it does not run contrary to it either.[11]

For the same reason that I just mentioned with regard to post-mortem directives, there is no compelling argument against respecting advance directives in these cases. As critical interests do not lose their authority simply because they cease to exist, nothing speaks against respecting the former wishes of the competent person as long as the demented patient does not exhibit any signs of an interest that stands in opposition to his former critical interests. If the patient's current interests align with her former critical interests or at least do not run counter to them, we have to respect her former critical interests to the extent that we have to respect her autonomy. Therefore, in the absence of any evidence to the contrary, we ought to honor advance directives that are expressive of these critical interests. Nevertheless, it would be wrong to say that the patient's former critical interests dominate her experiential interests. There is no conflict between these two kinds of interests in the first place, and in the absence of such a conflict, the question whether we ought to subordinate experiential interests to critical ones does not arise.

(B) *Conflict-cases*: It may be the case that there is a conflict between critical and experiential interests. Two kinds of conflict are possible: the colliding interests may exist contemporaneously or they may exist at different times.

(B1) *Simultaneous conflict*: In cases of a simultaneous conflict, critical and experiential interests exist at the same time. Leading a healthy life may be central to a person's conception of a good life, but while having a critical interest in living healthily she may also have an experiential interest connected to a strong desire to smoke a cigarette. Someone may have a critical interest in being a faithful spouse but also feel strongly tempted to commit an act of adultery and have an experiential interest in doing so. Or someone may have a critical interest in promoting science even at the expense of her personal well-being while having an experiential interest in taking an extended holiday rather than working.

---

[11] For a more detailed description of these cases, see Menzel and Steinbock (2013, 493), Menzel and Chandler-Cramer (2014, 25, 27).

In these cases of a contemporaneous conflict between critical and experiential interests, the authority of critical interests remains untouched by contravening experiential interests. The reason is that critical interests are (in part) *defined* as interests whose function is to control experiential interests. Being normative in nature, they are interests concerning what we ought to do, not interests in pleasurable experiences. To have a critical interest in promoting science means, among other things, to have a (critical) interest in (sometimes) not fulfilling those experiential interests that run contrary it.[12] Therefore, the normative authority of critical interests is not negated by concomitant, but contravening experiential interests. (To say that critical interests retain their normative authority is, of course, not to say that they will necessarily be action-guiding, for cases of weakness of will are always possible.) Since critical interests are those a person thinks she ought to act upon even in the light of contravening experiential interests, it is analytically true that to have a critical interest means to have an interest in not following (some) experiential interests.

(B2) *Non-simultaneous conflict*: There may, however, also be a non-simultaneous conflict between critical and experiential interests: for example, a person may lose her ability to have critical interests, but still have experiential interests, and these latter interests may run contrary to the critical interests the person had when she was a competent person (but no longer has now). At an earlier point of time t_1, a person P may have a critical interest that at a later point of time t_2 action A be performed, while at t_2, P, having lost competence, no longer has any critical interest, but still has an experiential interest, namely the experiential interest that A *not* be performed. These kinds of conflict occur when the experiential interests of a demented patient run contrary to the critical interests that she expressed in the advance directive as a competent person. In these cases, it is no longer analytically true that critical interests are more authoritative than experiential ones. Whereas to actually *have* a critical interest means to have an interest that is designed to control our experiential interests, *having had* a critical interest that is in conflict with one's current desires and wishes does not necessarily imply the normative authority of the now lost critical interests over the still existent experiential interests. Do critical interests dominate experiential interests in these cases? To answer this question, another distinction is called for, namely the distinction between a transitory and a permanent loss of competence.

(B2a) *Transitory loss of competence*: A first stage in which a person is competent may be followed by a second stage in which this person loses her competence and then by a third stage in which the person regains her competence. At t_1, a competent person may have the critical interest to lead a healthy and normal life. At t_2, under the influence of severe psychosis or severe depression, she may, deprived of her cognitive capacities, have the experiential interest in injuring herself. At t_3, having regained her competence, she may have a critical interest similar to the one she had at t_1 and therefore be grateful that the experiential interest she had at t_2 was not fulfilled.

---

[12]See Buller (2015, 702).

A famous example of this constellation is the case of Ulysses: Ulysses, at t_1, has a critical interest in preserving his life and in *not* following the death-bringing chant of the Sirens (which we may take as a metaphor for a temporary psychosis). He therefore orders his mates to tie him to the mast and to ignore his later pleas to release him. At t_2, Ulysses has an experiential interest in following the chant of the Sirens, but, being tied to the mast by his companions, is prevented from fulfilling it. At t_3, Ulysses, now no longer within the reach of the Sirens, regains his critical interest in preserving his life and therefore retrospectively welcomes that at t_2 his mates hindered him from following the chant of the Sirens. Thus, at t_2, Ulysses, having lost his ability to form critical interests, has an experiential interest that runs counter to his former critical interest, while at t_3, having regained his ability to form critical interests, he has a critical interest similar to the one he had at t_1.[13]

Someadvance directives in dementia have the structure of Ulysses-instructions. More precisely, this characterization applies to advance directives that relate to the early onset stages of dementia, most notably to stage 2 according to the "global deterioration scale" ("very mild cognitive decline"), in which the patient complains of memory deficits but is still clearly a competent person and shows appropriate concern about the symptoms of the disease. Imagine that a fully competent person, having undergone presymptomatic testing which indicates the onset of the disease before its symptoms are manifest,[14] anticipates that in the "phase of forgetfulness" she will show the first symptoms of dementia. She may fear that in this stage she will fail to perform certain tasks, that she will become aggressive because of a loss of impulse control, or that she will misplace an object of value. She may then instruct others to prevent her, if necessary by force, from doing harm in one of the anticipated ways. For example, she may instruct them to place the object of value beyond her reach or to administer, even by force, sedative drugs if she shows signs of aggressive behavior. Later, in a phase of temporarily receding lucidity in which she regains her impulse control and competence, she may welcome that someone hindered her from following her aggressive impulses or from misplacing the object of value in the preceding phase of incompetence.

With regard to these cases, we are inclined to judge that it is morally legitimate, perhaps even morally required, to disregard the demented patient's current experiential interests that run counter to her former critical interests. We usually find it morally unobjectionable and legitimate to override a psychotic patient's interest to injure herself, Ulysses' interest to be released from the mast or the early stage demented patient's impulse to be aggressive or to misplace an object of value. Does this mean that in these cases critical interests dominate later experiential interests?

The answer is in the negative. It is true that, when disregarding the patient's experiential interests at t_2, we act *in accordance* with the critical interest he had at t_1. But acting in accordance with critical interests does not mean that these critical interests

[13]For a more detailed discussion of the Ulysses-scenarios, see Dresser (1984), Quante (2002, 287–290), Davis (2009, 288–289), Elster (2010), Hallich (2011).

[14]For presymptomatic testing of dementia by means of blood tests and biomarkers, see Davis (2014, esp. 547–548).

are the *justificatory reason* for why we act as we do. Rather, many reasons speak in favour of the view that the justificatory reason for disregarding the experiential interest at t_2 is the subsequent consent of the person who has regained her competence.[15] As an analogous example, think of the paternalization of children: we often act paternalistically towards our children when it concerns their later welfare. That is, we disregard their current experiential interests (e.g. their experiential interest to eat ten ice creams and six pieces of candy a day) and act against them in order to promote their well-being. However, we do so not because we respect their preceding preferences—normally, our children do not sign consent forms on Monday authorizing us to disregard their food-preferences on Tuesday—but because we anticipate that they will later retrospectively consent to this treatment. Paternalization of children shows the possibility of legitimately overriding a person's interests in the absence of a directive that explicitly authorizes us to disregard a subject's wishes and interests at a later point of time.

Thus, we should not invoke Ulysses-cases to support the dominance thesis. Even if we think (as we usually do) that we are justified in disregarding a subject's experiential interests that run contrary to preceding critical interests in cases of a transitory loss of competence, this view does not commit us to accepting that we do so because we respect these preceding interests. It does not imply that these critical interests override experiential interests.

(B2b) *Permanent loss of competence*: The loss of competence in cases of a non-simultaneous conflict between critical and experiential interests may also be permanent. The person who has lost her competence may not regain it. This may occur for one of two reasons. The first reason is that the person is in an advanced stage of the disease, more specifically in stage 6 ("severe cognitive decline") or 7 ("very severe cognitive decline") according to the "global deterioration scale". In these stages, the patient is largely unaware of recent events and experiences and of her surroundings. Moreover, she may have lost all verbal abilities and psychomotor skills. Her cognitive capacities have been irrevocably lost. The second possible reason for a permanent loss of competence is the enactment of an advance directive that requires ending the patient's life in a state of incompetence. If a directive—even one that pertains to an earlier stage of dementia in which the patient has only temporarily lost her competence but may still regain it—demands the ending of the patient's life in a state in which she has lost her cognitive skills, the loss of competence *turns out* to be permanent if the directive is enacted. Assume that at t_1, a competent person expresses the wish that, should she be in an early state of dementia at a later point of time t_2, life-sustaining treatment ought to be forgone in, e.g., the case of pneumonia. At t_2, the patient, now being in an early stage of dementia with a temporary loss of competence, appears to take delight in basic pleasures such as eating or quietly sitting in the sunshine. Should she catch pneumonia, fulfilling her current wishes would require keeping her alive, so it would mean ignoring the express wishes of the former competent person. By contrast, ignoring his current wishes and honoring the

[15] For a detailed defence of this view, see Hallich (2011, 159–165). For a discussion of the role of subsequent consent, see also Davis (2004).

directive would mean bringing about her death. As death brings with it a permanent loss of competence, the loss of competence would then *turn out* to be permanent in the light of the enactment of the directive. This holds true notwithstanding the fact that, without the enactment of the directive, the loss of competence would have been only transitory.

In these cases, unlike in Ulysses-cases, it is not an option to justify the disregarding of the patient's experiential interests with reference to subsequent consent. As late dementia or the enactment of an advance directive that requires the termination of the patient's life will be followed by death, not by a state in which competence is regained, we cannot claim that the justificatory reason for disregarding a demented person's interests is that this person will later, in a state of regained competence, consent to it. Rather, a defender of the view that the principle of precedent autonomy authorizes us to act against the demented patient's wishes in these cases has to show that critical interests dominate experiential interests even if the patient has *permanently* lost her cognitive skills (or will lose them if the directive is enacted). In what follows, I will attempt to show that this claim is false. In these cases, or so I argue, experiential interests negate the authority of critical interests.

10.4 Can Experiential Interests Negate the Authority of Critical Interests?

Critical interests *possess* normative authority insofar as they count as interests that we have to respect to the extent that we have to respect a person's right to autonomy. Accordingly, they lose their normative authority if they cease to have this status, i.e. if they no longer define the set of wishes and desires that we have to honor to the extent that we have to respect a person's autonomy. (Of course, autonomy can be outweighed by other considerations when it comes to decide what we ought to do all things considered. Therefore, critical interests generate only prima facie obligations to respect someone's will.)

How, then, can critical interests be deprived of their normative authority? Examples of a loss of authority of critical interests come to mind easily. Think of a deeply religious person who, in the course of her life, becomes an atheist. At the age of twenty, she defines her goals in life in terms of pleasing God and leading a pure and spiritual life, and she is firmly dedicated to pursuing the ideal of a pious and solitary existence. At forty, she has an interest in worldly pleasures, in promoting her own welfare and in having a family. At twenty, she believes in the sanctity of human life and is convinced that human lives ought to be prolonged at all costs, while at forty, she has come to believe that the prolongation of a life full of pain and dependence on others ought to be avoided. The plans and projects that shaped her life when she was twenty have undergone a complete revision by the time she is forty.

In this case, the former set of preferences has lost its normative authority in the later stages of the person's life. An advance directive that stems from the religious beliefs and preferences the person had at twenty would no longer be expressive of preferences and wishes that possess normative authority when the person is forty. It would therefore have lost its validity. It is important to be clear about the sense of "valid" that is at stake here: there is *a* sense of "valid"—let us call it the "ethical" sense—in which it is correct to say that an advance directive loses its validity if the preferences to which it gives expression cease to be authoritative. This loss of validity does not preclude the possibility that the directive is still *legally* valid. One can easily imagine advance directives that are legally binding even though they no longer represent the person's current wishes. In this case, the person who made the directive is comparable to someone living in an unhappy and ruined marriage that is still a marriage according to the law: her advance directive has lost its ethical validity, but it is still valid "according to the law". When we talk of advance directives ceasing to be valid because one set of preferences loses its normative authority, we draw on the ethical, not on the legal sense of "valid".

Examples like the one of the believer who becomes an atheist help us bring to light the conditions that have and, even more importantly, those that do not have to be satisfied in order for critical interests to lose their normative authority. What happens in the above example is clear: a set of critical interests is, in the course of time, replaced by another and radically different set of critical interests. Unlike in the *no-conflict-scenarios* of the type (A2) and (A3), critical interests here do not just cease to exist—in which case, as I argued above, they would retain their normative authority—but are rescinded on account of a different set of critical interests that is incompatible with the first. "Incompatible" means that a person could not coherently identify with these two sets of preferences at the same time: one cannot be a deeply religious person while also being a convinced atheist and one cannot have a life plan to lead a solitary life and a life plan to have a family at the same time. Thus, a person's earlier preferences no longer constitute the set of preferences we have to honor in order to respect her autonomy if another set of preferences that cannot coexist with the first set has eradicated them. This condition of the replacement of one set of preferences by another set that is incompatible with the first—call it the *replacement-condition*—is a *necessary* condition for critical interests to lose their normative authority and consequently for the invalidation of an advance directive that is expressive of these critical interests. In Sect. 10.4 I will defend the claim that this replacement may invalidate the advance directive even in cases in which the set of critical interests is not replaced with another set of critical preferences but "only" with experiential preferences.

In cases of anon-simultaneous conflict between critical and experiential interests, the replacement condition is satisfied. Think, for example, of Dworkin's famous "Margo case"[16]: Margo is an intellectually agile person. Dworkin invites us to

[16]See Dworkin (1994, 220–229). The Margo-case and Dworkin's normative assessment of the case have become a constant reference point in the literature on advance directives in dementia. It has elicited Dresser's trenchant criticism (see, e.g., Dresser 1995); for further comments on the Margo-case, see, e.g., Jaworska (1999), McMahan (2002, 498–502), Sumner (2011, 108–117), Menzel and Steinbock (2013, 489–492), Beauchamp and Childress (2013, 228–229), Hawkins (2014, esp. 510–519).

imagine that in a state of competence, she expresses revulsion at the prospect of becoming demented and being dependent on others. She firmly expresses (perhaps by means of an advance directive) her desire that, should she ever become demented, she would no longer want to be kept alive if she, for example, got pneumonia. Margo then becomes demented, but still seems to be, according to a medical student who regularly visits her, "one of the happiest people I have ever known",[17] who enjoys quietly sitting in the sunshine, caressing animals, or eating peanut-butter-and-jelly sandwiches and who exhibits all signs of joy in life. The demented Margo evidently has an interest to live that runs contrary to her former critical interests. The replacement-condition is satisfied: the original set of critical preferences is replaced by another set of preferences.[18] This latter set of preferences stems from experiential interests that are incompatible with the first: it is impossible to honor the competent person's wish that life-sustaining treatment be withheld in a state of dementia and the demented person's interest to live at the same time.

It is also important to highlight the conditions that do *not* have to be satisfied for critical interests to lose their normative authority. First, the loss of authority of critical interests does not require a *decision* to change one's preferences (Let us call this condition the *decision-condition*.). The above example illustrates that critical interests may lose their authority as the result of a gradual process that may stretch over a considerable span of time. It will often be impossible to mark a clear turning point that indicates the point of time at which the new preferences were "born": at forty, the person may not be able to say when exactly she lost her faith. It may be just as impossible for her to say when she became a disbeliever as it is impossible for someone to fix the point of time at which her hair turned grey. She may never have willingly made a decision to abandon her faith. There may also be overlapping phases in which the person is still partially religious but has already to a great extent lost her religious beliefs. The loss of normative authority of critical interests may happen by and by.

The impact of this observation on advance directives in dementia is evident: The mere fact that a demented patient has not decided to abandon the preferences that she expressed in the advance directive does not guarantee that these preferences retain their normative authority. It does not rule out the possibility that the advance directive that gives expression to these preferences has become invalid by other means.

Finally, in order to lose its validity, an advance directive need not be *revoked*.[19] (Let us call this condition the *revocation-condition*.) A revocation is an overt act by means of which a person gives expression to the withdrawal of a former decision.

---

[17] Dworkin (1994, 221).

[18] For the time being, I ignore the fact that in this case, the original set of critical interests is replaced not by another set of critical interests, but by experiential interests. I address this issue in Sect. 10.4, where I will argue that the answer to the question of whether critical interests lose their normative authority or not does not hinge on whether they are replaced by critical or by experiential interests.

[19] Pace, e.g., King (2007, 204–213). According to King, a revocation is the only way of invalidating an advance directive. This claim mirrors the rather loose use of the term "revocation" in the literature on advance directives. Often, the term is used in a very general sense to cover *all* forms of invalidation of an advance directive.

For example, someone might tear apart a consent form that authorizes others to withhold food and water in late-stage dementia, thereby revoking her former consent. *A revocation is a sufficient but not a necessary condition for an advance directive to lose its validity.* That is, whenever a person revokes an advance directive it ceases to be valid. Nevertheless, it is also possible that the advance directive loses its validity without being revoked because it is possible that the preferences expressed in an advance directive lose their normative authority without being overtly taken back or in any way explicitly disaffirmed. Clearly, an advance directive loses its validity without being revoked if, after a certain time span, it has not been explicitly disaffirmed, but not explicitly renewed either. If a considerable time span has elapsed since the directive has been made and it has not been reaffirmed or renewed since then, we are likely to regard this as sufficient evidence that the advance directive no longer reflects a person's present set of preferences. We would therefore decline to regard it as binding notwithstanding the fact that the patient did not revoke it.

Again, this observation directly bears on advance directives in dementia: since revocation is not a necessary condition for an advance directive to lose its validity, an advance directive in dementia may lose its validity even though it has not been explicitly revoked. In other words, the mere fact that the demented patient has not revoked the advance directive does not guarantee its validity.

In cases of a non-contemporaneous conflict between critical and experiential interests and a permanent loss of competence, the necessary condition for the invalidation of an advance directive, the replacement condition, is fulfilled, while the revocation-condition and the decision-condition are not fulfilled. However, since the latter two conditions are sufficient, but not necessary conditions for the invalidation of advance directives, the fact that they are not satisfied does not call into question the view that in cases of this kind, experiential interests negate the normative authority of critical interests and invalidate the advance directive. Rather, a preliminary conclusion is as follows: if the basic desires and wishes of the demented patient run contrary to the critical interests of the once competent person, these wishes and desires take over the normative authority of the former critical interests. They then form the set of interests that we have to respect to the extent that we have to respect the person's autonomy. Should the "happy demented" like Dworkin's Margo exhibit all signs of a desire to live, then this desire, stemming from her experiential interests like the interest in enjoying good food or sitting quietly in the sunshine, outweighs the former critical interest that life-sustaining treatment ought to be withheld in a state of dementia. Insofar as the right to autonomy is concerned, this desire, so my tentative conclusion goes, has to be respected notwithstanding the fact that the competent Margo articulated the express wish that it be ignored.[20]

[20] Admittedly, whether the principle of autonomy requires the satisfaction of experiential interests or not is a contested issue. On a broadly Kantian model, one might argue that it is only interests that are based in reason, i.e. critical interests that express our autonomy and that are protected by the right to autonomy, whereas experiential interests based on pleasure or desire are not. However, the main idea of my argument does not hinge on a positive answer to the question of whether the principle of autonomy requires us to respect experiential interests or not. Even if the answer to this question is in the negative, the fact remains that the experiential interests of the demented patient *are* interests

10.5 Three Objections

Of course, this cannot be the end of the story. The foregoing argument is likely to elicit the objection that it simply ignores the crucial distinction between critical and experiential interests. It appears to sidestep the question of whether the dominance thesis is true or not by focusing on cases in which the difference between critical and experiential interests is irrelevant. If a religious person becomes an atheist in the course of her life, critical interests lose their authority because other *critical* interests eradicate them. Things seem to look differently if it is *experiential* interests that replace critical interests. Why, for example, should a demented patient's pleasure in good food that gives rise to an experiential interest be on par with the former critical interest of an intellectually agile person that in a state of dementia, life-sustaining treatment ought to be withheld? In what follows I will examine three lines of argument to support the view that in scenarios of the kind that are now under consideration critical interests are *unlike* experiential interests in that they alone may deprive our former critical interests of their normative authority. All of them, or so I argue, are unpersuasive.

10.5.1 Asymmetry of Cognitive Capacities

The first line of argument draws our attention to the fact that there is an asymmetry between the cognitive capacities of the competent person who wrote the advance directive and the demented person who has lost all cognitive skills. This asymmetry, so the objection goes, speaks against the view that preferences that originate from experiential interests may negate preferences that stem from critical interests.[21]

It goes without saying that the asymmetry the objection points to really exists. However, it need not disconcert us. In various contexts, we often place more rigid standards of competence on validly assenting to a procedure than on validly withdrawing from it. For example, in order for a subject to validly consent to participation in a cancer research program, this subject usually has to meet criteria of informed consent, i.e. consenting in light of a full understanding of the procedure to which she is about to submit herself, an understanding of the relevant information, an awareness of the risk her participation implies etc. In order to take back his consent validly, however, he need not fulfill these conditions: if, in a state of restricted competence, e.g. under the influence of misinformation, he tears apart the consent form, we will

---

that—in the absence of compelling arguments to the contrary—deserve to be respected. One might plausibly argue that a cat's interest not to be exposed to boiling water ought to be respected without invoking the principle of autonomy to ground this moral obligation. Analogously, as long as the principle of precedent autonomy turns out to be unpersuasive in conflict cases of the kind now under consideration, there is good reason to assume that the experiential interests of the demented patient, be their satisfaction required by the principle of autonomy or not, ought to be honored.

[21] See, e.g., King (2007, 204–213). King endorses the view that a valid revocation must exhibit a degree of autonomy similar to the one that is required to make the advance directive valid.

accept this as a valid revision of his former decision. The justification for this asymmetry is as follows: higher standards of competence for entering a procedure shield the subject against precipitately taking part in a procedure which may involve serious dangers and risks. Conversely, lower standards of competence for stepping back from it are justified because by stepping back from it the subject merely reenters into the "default position". Under normal circumstances, there is no risk in *not* taking part in a medical experiment. Therefore, the competence requirements for repealing one's consent to participate in a research program are lower than those a subject has to meet for validly entering into it.[22]

Similarly, the competence standards that have to be met in order to *in*validate an advance directive are less rigid than those that have to be met for making a valid advance directive. Someone will not succeed in making a valid advance directive if, on a whim, she formulates her end-of-life directive that artificial nourishment be withheld in a terminal state of a disease. By contrast, if she tears apart the advance directive (and does not renew it) in a freakish or whimsical state of mind, this will be accepted as an invalidation (in this case even as a revocation) of the advance directive. Making a valid advance directive usually requires that the conditions of informed consent be satisfied, i.e. that the person who makes the directive has a correct understanding of the relevant information, that her reasoning faculties are intact and that she exercises them when making the decision.[23] Invalidating the directive does *not* require that these conditions be satisfied.

As in the case of consent to participation in a research program, this asymmetry is well justified. Different risks are at stake in making and in repealing an advance directive. By making an advance directive, a person makes a risky decision insofar as she determines her future treatment without being able to completely anticipate the future state of affairs to which her directive relates. Although we usually regard advance directives as a means of reducing the risk of being exposed to life-prolonging measures against our will at the end of our lives, the fact remains that we can never know exactly whether our future intentions and desires will match with the will that guided our decisions when we made the advance directive. It is always uncertain and risky to anticipate one's future will. By contrast, when withdrawing the directive, the subject reenters into the "default position", i.e. a state in which she leaves open how she will be treated in the future. As making an advance directive is, in this sense, a risky decision, whereas repealing it is not, it is reasonable that the standards imposed

[22]In a similar vein, Buchanan and Brock defend a "sliding-scale account" of competence (1990, 17–86 [Chap. 1]). They argue that "no single standard of decision-making competence is adequate" (ibd., 52) and that "the standard of competence ought to vary in part with the expected harms or benefits to the patient of acting in accordance with the patient's choice" (ibd., 51). As Beauchamp and Childress have noted, we should not take this claim to mean that competence itself varies in relation to risk, since risky decisions do not require more ability at decision making than less risky ones. Rather, the core idea of the "sliding-scale account" should be that the *evidence* for determining competence varies according to risk; see Beauchamp and Childress (2013, 119–120).

[23]For a standard account of informed consent, see Beauchamp and Faden (1986, 274–297) (Chap. 8), Beauchamp and Childress (2013, 120–140).

on validly making an advance directive are more demanding than those imposed on validly withdrawing it.

Hence, the mere fact that a demented patient, having lost competence, is unable to comply with the standards of competence that have to be met for an advance directive to be valid does not imply that the patient is unable to conform to the standards that have to be met to invalidate the advance directive. The asymmetry between the cognitive skills of the competent person and those of the demented patient, undeniable as it is, is not at odds with the view that experiential interests may negate the normative authority of critical interests.

10.5.2 Anticipation of Future Preference Changes

A second objection to the view that experiential interests of the demented patient may negate the authority of contravening critical interests of the once competent patient is the following one. When writing an advance directive, the competent patient *foresees* that, when being in a state of dementia, she will have the desire that her physicians and relatives disregard her former critical interests and act upon her later experiential interests. Therefore, so the objection goes, not fulfilling these later experiential interests is exactly the *point* of advance directives in dementia.[24] To bring out the salient point of the objection more clearly, promises provide an illuminating analogy: assume that, when marrying, I promise my wife that I will not betray her. I may, when marrying, foresee that I will later be exposed to temptations in which I will have a strong inclination to become unfaithful. At a later point of time, I may actually be exposed to these temptations. But since not betraying my wife is exactly what I have promised her, it would be absurd to say that my later desire to betray her negates the promise I made to her. The point of the promise is just to make sure that I will not follow my desire; therefore, my having a desire to do what I promised *not* to do does not suspend or negate its binding force. One could argue that, for the same reason, experiential interests cannot negate the normative authority of critical interests in scenarios of the kind now under consideration. Just as not following my desires is what keeping my promise consists in, acting contrary to the demented patient's experiential interests is what respecting the person's autonomy consists in. As Feinberg memorably puts it, the right of autonomy is like the sovereignty of a nation over its

[24]Sumner states this point very clearly: "It would be perverse to disregard my earlier directives simply because I am now in the very condition of incapacity I anticipated in recording them" (2011, 111).

territory[25]: it extends over a *sphere* of interests. This sphere comprises those contravening interests that the competent patient who exercises her right to autonomy foresees.[26] Therefore, the objection goes, the principle of autonomy demands negating these foreseen preferences.

Though intriguing, this argument is flawed for at least two reasons. First, we have to take into consideration that we change not only our preferences, but also our evaluations of foreseen changes of preferences. A person who, when making an advance directive, foresees a future change of her preferences, may still, at a later point of time, come to reevaluate this foreseen change of preferences. Often, it will be impossible to decide whether a preference that runs contrary to the one the competent person expressed in the advance directive is exactly the one that the competent patient foresaw or whether it is indicative of a reevaluation of the foreseen change of preferences. If the second alternative obtains, the alteration of the patient's evaluation of the foreseen preference change must be respected as an exercise of personal autonomy.[27]

By way of illustration, the following example might be helpful. A smoker sincerely wants to overcome her addiction. She thinks that she ought to abstain from nicotine because leading a healthy life is more important to her than the fulfillment of short-lived hedonic desires. She foresees that she will soon have a strong desire to smoke a cigarette. She attempts to forestall the fulfillment of this desire by instructing all her friends to make sure that, as soon as the desire arises, she will have no access to nicotine. Later, as was foreseen by her, she has a strong desire to smoke a cigarette. How can we decide whether the desire to smoke a cigarette is exactly the wish that she foresaw and whose fulfillment she tried to forestall or whether it is expressive of a different evaluation of the preference change she foresaw? It seems impossible to exclude the second alternative. The desire to smoke a cigarette may be the result of a revision of the former view that leading a healthy life is more important to the person's life plan than the fulfillment of short-lived desires. For example, the smoker

---

[25]Feinberg (1986, 47–51).

[26]Davis (2004, 273–281, 2009, 365–367) labels this view the "extension view" of autonomy and plausibly ascribes it to, e.g., Rhoden and Dworkin.

[27]This problem vitiates Davis' attempt to base precedent autonomy on a "resolution preference", i.e. an evaluative judgement that is formed in awareness of two conflicting preferences and that is either contemporary with the two conflicting preferences or, if these exist at different times, with one of them, typically the former one. (2004, 273–281, 2009, 365–367). According to Davis, a resolution preference expresses what a person wants, all things considered. Therefore, to respect his or her autonomy, we must respect his or her resolution preference. However, even resolution preferences may be subject to changes of mind; a person may come to abandon or to alter his or her resolution preference. Similar problems arise with King's suggestion to incorporate an "antirevocation provision" in advance directives in order to protect the subject against involuntary revocations of his former will (King 2007, 204–213). This antirevocation condition, King suggests, should specify circumstances under which the apparent revocation of an advance directive should not be honored. But again, even the antirevocation provision is prone to changes of will and may therefore become invalid. For the problem of anticipated preference changes, see also Kleinig (1984, 57), who rightly states: "consent is not to be construed after the manner of an irrevocable contract but as an ongoing commitment that may be reassessed and given up" (57).

may have come to believe that preserving her health is not that important to her life plan after all or that all the medical reports she read about the lethal consequences of smoking were unfounded. If that is the case, the altered evaluation of the foreseen preference change has to be honored as an expression of the subject's autonomy. Since there is no compelling argument to show that it cannot be the case, the mere fact that the preference change was foreseen does not warrant the claim that the first of the two conflicting preferences ought to be acted upon.

The same problem occurs with advance directives in dementia. A person may issue an advance directive in which she instructs her physicians to forgo life-sustaining treatment when she has become demented. She may foresee that, when being demented, she will strongly want her physicians to disregard her former critical interest and that, contrary to her express wish, she will then demand that life-sustaining treatment be provided. As a demented patient, she may really come to have the wish that life-sustaining treatment be provided. Still, it remains an open question how the demented patient's preference has to be interpreted: is it exactly the preference the competent person foresaw and whose fulfillment she tried to prevent or is it a new preference that gives expression to the reevaluation of the previous wish that the later foreseen preference not be satisfied? Again, as we not only change our preferences, but also our evaluations of foreseen changes of preferences, we cannot rule out the possibility that the later preference indicates that the patient has reevaluated the preference change she (rightly) foresaw as a competent person. If this is the case, we have to respect this reevaluation as an exercise of personal autonomy.[28]

Second, the possibility of foreseeing a future change of preferences is not specific to the relation between critical and experiential interests. It is true that, as in the case of an advance directive in dementia, someone may have a critical interest that her foreseen contravening experiential interests be negated. But it is also true that someone may have a critical interest that her foreseen future *critical* interests be negated. In this case, however, the later critical interests do *not* lose their authority to negate the former critical interests although these later interests are those whose fulfillment the person who had the former critical interests tried to prevent. Think of a profoundly religious person who is devoted to leading a pious and pure ascetic life and wishes to bequeath all her property to a religious organization. She may foresee that one day under the influence of a secular environment she will lose her faith. She may try to guard herself against this foreseen change of her (critical)

[28]This holds true especially in view of the fact that demented patients do have the capacity to value, as Jaworska (1999) has convincingly argued. Referring to Dworkin's distinction between critical and experiential interests, Jaworska points out that this distinction is too coarse and needs to be replaced by a more fine-grained one. She categorizes the valuations of demented patients as critical interests, thereby introducing a sense of "critical interest" that is different from the one that Dworkin draws upon. Whether we follow her in doing so or not seems to me a question of minor terminological importance. Jaworska's crucial insight is that the preferences of demented patients are distinct from mere desires, that they may, at least in the first stages of dementia, still be stable and that, although they do not imply grasp of one's life as a whole, they are expressive of valuations. As valuing involves reflection, the capacity to value, so Jaworska argues, shows that demented patients have not lost their autonomy. For Jaworska's critique of Dworkin's distinction between critical and experiential interests, see also Menzel and Steinbock (2013, 490–492).

interests and may time and again confirm her last will and her intention to donate her property to the religious organization. Nevertheless, one day it may happen that she actually loses her faith and becomes an atheist. If, in these new circumstances, she distances herself from her former decision and makes clear that she no longer wishes to bequeath her property to the religious organization, this changed preference will be respected notwithstanding the fact that she foresaw and vainly tried to forestall it at an earlier point of time. So the foreseeability of the preference change does not diminish the normative authority of the later preferences. If this is the case, the fact that a person foresaw this change is of no relevance to the authority of her later interests, and this holds true even in cases in which no other critical interests, but experiential interests replace former critical interests.

10.5.3 Qualitative Difference Between Critical and Experiential Interests

A third objection to the view that experiential interests may negate the normative authority of critical interests draws on a value judgement about these two kinds of interests. The objector may contend that there is a *qualitative* difference between them. Critical interests, he might argue, are superior in worth to experiential interests. He could spell out the argument in various ways. First, he could argue that critical interests are superior in worth to experiential interests because they give expression to distinctively *human* capacities. (Animals, though they may have the capacity to form experiential interests, lack the capacity to form critical interests.)[29] Second, the objector might argue that critical interests are superior in worth to experiential interests because they give expression to what constitutes our lives as *persons*. If "critical interests", unlike experiential ones, define what it is to be a person, they define the set of preferences that we have to honor to the extent that we have to respect a person's autonomy.[30] Third, in order to justify her claim that critical interest are superior in worth to experiential interests, the objector may adduce the fact that most members of a society *regard* critical interests as superior to non-critical ones.[31] The

[29]Dworkin, though not explicitly, embraces this view when he grounds the dominance thesis in considerations concerning human dignity: "I have been arguing that we not only have, in common with all sensate creatures, experiential interests in the quality of our future experiences but also critical interests in the character and value of our lives as a whole. These critical interests are connected, as I said, to our convictions about the intrinsic value—the sanctity or inviolability—of our own lives. […] Dignity is a central aspect of […] the intrinsic importance of human life" (1994, 235–236).

[30]Quante defends this view: "To lead one's life as a person means to identify oneself, be it in a positive or in a negative way, with one's future, one's future experiential interests included. If a culture attaches central value to leading one's life as a person, there is a strong reason for acknowledging and respecting those valuations of a person that he or she expresses in an advance directive as an exercise of his or her personal autonomy" (2002, 286; transl. by author).

[31]See the quotation from Quante (2002) in the preceding footnote, which allows for this interpretation as well.

objection then points to the social fact that there is a prevalent attitude among most members of a society that in cases of a conflict critical interests ought to override experiential interests.

I cannot examine these arguments in detail. It suffices to say that much more argument would be needed to make them persuasive. As they stand, they appear to rest on a naturalistic fallacy. As is well known, we cannot infer value judgements from purely descriptive premises.[32] Thus, we cannot infer the statement that critical interests are superior in worth to experiential interests from the purely descriptive statement that the capacity to form critical interests is specific for human beings.[33] Similarly, the mere fact that critical interests constitute our lives as persons does not make them superior in worth to experiential interests. Of course, one is free to define "person" as an evaluative concept and to stipulate by definitional fiat that personal interests, *because* they define what a person is, are superior in worth to non-personal ones, thereby avoiding a naturalistic fallacy. However, the argument that critical interests are superior in worth to experiential interests *because* they define our lives as persons then becomes blatantly circular. It boils down to saying that critical interests are superior in worth to other interests because they are superior in worth to other interests. Finally, the mere social fact that most people in a society accept the view that critical interests ought to dominate experiential interests does not validate the value judgement that critical interests are superior in worth to experiential interests either. The majority of members of a society may accept the most absurd views. The mere social fact that people *regard* critical interests as superior to non-critical ones does not warrant the claim that they are right in doing so.

Even leaving these worries aside, the objector faces another problem. Let us, for the sake of the argument, assume that a value judgement about the superior worth of critical over experiential interests is true. Still, its truth does not imply the truth of the normative judgement that in cases of a conflict between the two kinds of interests we *ought* to respect the critical interests even at the price of disregarding the demented patient's current experiential interests. It is a common vice in moral philosophy to overlook the categorical difference between value-judgements and "ought"-judgements, particularly if both kinds of discourse are lumped together under headings such as "prescriptive language" or, even worse, "moral language". As G.H. von Wright pointed out most forcefully, the proposition that we ought to do what is good does not express a conceptual truth. Value concepts are not intrinsically normative.[34] Therefore, the transition from "This is good" to "This ought to be done" or to "This ought to be realized" is not a logical transition, and the transition from

---

[32]See, e.g., Hare (1952, 1–55, 79–93) (Chaps. 1–3, 5). It would be beyond the scope of this paper to go into the details of the discussion of metaethical naturalism. I simply take it for granted that metaethical naturalism cannot be consistently defended.

[33]For a classical critique of the "speciecism" that is involved here, see Singer (2011, 48–70) (Chap. 3).

[34]See von Wright (1963, 155–157). According to Moore's influential analysis of the relation between "good" and "ought", normative concepts like "ought" and "duty" can be analyzed in terms of "good", whereas "good" itself, being a simple concept, cannot be analyzed in terms of normative concepts; see Moore (1993, 76–78, 196–198) (Chap. 1, § 17, Chap. 5, § 89).

"Interest A is superior in value to interest B" to "In cases of a conflict between A and B, A ought to be respected" is not a logical transition either. The dominance thesis is a *normative* thesis; it is a claim about what kind of interests we *ought* to honor in cases of a conflict between critical and experiential interests. Being normative, it does not follow from a value judgement about the superior worth of critical over experiential interests.

As all three objections turn out to be unpersuasive, I see no reason to take back the preliminary conclusion that I argued for in the preceding section: in scenarios of a non-contemporaneous conflict between critical and experiential interests, experiential interests can negate the normative authority of critical interests.

10.6 Concluding Remarks: The Binding Force of Advance Directives in Dementia

The upshot of the argument is as follows: if the experiential interests of the demented patient run contrary to the critical interests of the once competent person that were expressed in the advance directive, the critical interests do *not* dominate the experiential ones. Therefore, the principle of respect for autonomy does *not* commit us to subordinating the interests of the demented patient to those of the competent person. Thus, experiential interests may override critical interests.

To guard against misunderstandings, I wish to emphasize that this conclusion is not to deny that advance directives are a highly useful tool for guiding medical choices. They are far from being superfluous. This is so for two reasons.

First, the view that experiential interests may negate the authority of critical interests applies only to (some) cases in which these two kinds of interests *collide*. In the absence of a collision of this kind, critical interests retain their authority. As I argued at length with regard to "no-conflict scenarios", the mere fact that critical interests cease to exist does not deprive them of their normative authority. Their authority is also left untouched in cases of a simultaneous conflict between critical and experiential interests. Furthermore, in "Ulysses-cases" we are justified in acting in accordance with the former critical interests as long as we can reasonably expect the patient's subsequent consent. In sum, should the demented patient not exhibit any signs of a contravening interest, there is no argument against respecting the former wishes of the competent person with regard to how she is to be treated in a state in which competence has been lost.[35]

[35] A similar point is made by McMahan (2002, 497–498), Menzel and Steinbock (2013, 492–493), Menzel and Chandler-Cramer (2014, 24–29). Rhoden neglects this important point. She contends that rejecting precedent autonomy implies that the competent patient's right to refuse treatment would be upheld only as long as she remains competent. "Taken to an extreme, this could mean that a Jehovah's Witness could refuse a blood transfusion until he "bled out" and became incompetent, after which he could be transfused" (1990, 857). However, this counterintuitive consequence follows only if the rejection of the principle of precedent autonomy is extended to no-conflict cases, which it need not be.

Second, in cases of a transitory loss of competence, i.e. in cases in which the person regains her competence after a phase in which she is unable to make competent decisions, advance directives help us to rightly anticipate a person's subsequent consent to a treatment she has been subjected to during a phase of incompetence. For example, a person who anticipates that under the influence of a severe psychosis she will want to commit suicide may write an advance directive that authorizes others to prevent her, if necessary by force, from committing suicide in this state of a loss of competence. In this case, the advance directive corroborates the assumption that the patient will most likely consent to having been prevented from committing suicide by force after the phase of incompetence. Similarly, advance directives that pertain to the onset stages of dementia will increase the reliability of judgements about the patient's later consent to a treatment during a preceding phase of a loss of cognitive skills that she is expected to give in a later phase of recurring competence. The directive may, for example, increase the reliability of the judgement that the demented person will, in a phase in which she temporarily regains her competence, consent to having been hindered by force from misplacing an object of value in a preceding phase of the disease in which she had lost her competence. Thus, insofar as subsequent consent is relevant to justifying a treatment during a phase of a loss of competence, advance directives are an important tool for guiding medical choices because they make judgements about an expected subsequent consent more reliable than they would be without an advance directive.

Bearing these two caveats in mind, can we conclude that in cases of a non-simultaneous conflict between critical and experiential interests and a permanent loss of competence we ought to disregard the critical interests and act in accordance with the experiential interests? Perhaps surprisingly, the answer is "No". The foregoing argument does *not* rule out the possibility that we should disregard the patient's actual experiential preferences and act in accordance with the competent patient's wishes. It only rules out the possibility that we justify doing so by drawing on the principle of precedent autonomy. There is nothing contradictory in claiming that we are justified in acting *in accordance* with the former interests without resorting to the principle of precedent autonomy, i.e. without claiming that the reason for doing so is that we respect or honor these preceding critical interests. (As I stressed above, "to act in accordance with the former interests" does not mean that these former interests are necessarily the justificatory reasons for acting in accordance with them.)[36] For example, within a utilitarian framework, one might point to the fact that there is a strong societal interest in giving critical interests priority over

[36]E.g., McMahan's sagacious defence of the view that the competent person's preferences ought to be accorded more weight than those of the demented patient is independent of the principle of precedent autonomy. McMahan construes conflict cases as conflicts between two distinct parts of a life, namely between the earlier good of the competent self and that of the later incompetent self. While it may be in the best interest of the demented patient to live, ignoring the directive may retroactively diminish the value of the life prior to the onset of dementia. McMahan argues that physicians should give priority to the earlier part of the life because it is overwhelmingly the dominant part of the life as a whole (2002, 498–503). For a detailed discussion of McMahan's position, see Hawkins (2014, 514–515).

experiential interests, i.e. that most members of a society have a strong interest in subordinating experience-related desires and wishes to critical interests in cases of a conflict. Impartially universalizing these interests, one might argue, will lead us to prioritize critical over experiential interests because doing so will, in sum and considered from an impartial point of view, maximize the preference-satisfaction of all parties involved.

The price to be paid for pursuing this line of argument would be that we openly admit that acting against the demented patient's wishes is not an instance of respecting her precedent autonomy but rather an instance of privileging society's (or the majority of its members') interests. It would require abandoning the pleasant and flattering view that, when acting in accordance with the wishes the patient articulated in the advance directive, we do so for the sake of the patient. It would compel us to accept the less flattering view that when placing critical interests over experiential interests we are not good Samaritans selflessly protecting the patient's autonomy but are simply trying to enforce the implementation of decision-making standards that most members of a society approve of.[37] If, by contrast, we stipulate that our decision should be guided by the attempt to respect the patient's autonomy, then, contrary to what is often thought, we lack arguments to respect the competent person's will in cases in which the later experiential interests run contrary to the foregoing critical interests. We should then disregard the advance directive.

References

Alzheimer's Association. 2015. Alzheimer's Disease Facts and Figures. *Alzheimer's & Dementia* 2015. https://www.alzheimers.org.uk/site/scripts/documents_info.php?documentID=412. Accessed October 11, 2015.

Alzheimer's Society. 2015. Dementia 2014 report. Statistics. https://www.alz.co.uk/research/statistics. Accessed October 11, 2015.

Beauchamp, Tom, and Ruth Faden. 1986. *A History and Theory of Informed Consent*. Oxford: Oxford University Press.

Beauchamp, Tom, and James Childress. 2013. *Principles of Biomedical Ethics*, 7th ed. Oxford: Oxford University Press.

Buchanan, Allen. 1988. Advance Directives and the Personal Identity Problem. *Philosophy & Public Affairs* 17: 277–302.

Buchanan, Allen, and Dan Brock. 1990. *Deciding for Others. The Ethics of Surrogate Decision Making*. Cambridge: Cambridge University Press.

Buller, Tom. 2015. Advance Consent, Critical Interests and Dementia Research. *Journal of Medical Ethics* 41: 701–707.

Cantor, Norman. 1992. Prospective Autonomy: On the Limits of Shaping One's Postcompetent Medical Fate. *Journal of Contemporary Health Law & Policy* 8: 13–48.

Cantor, Norman. 2005. *Making Medical Decisions for the Profoundly Disabled*. Boston, MA: MIT Press.

[37] In the same spirit, Dresser and Robertson object that it is fundamentally dishonest to cloak those value-judgements that underlie the subordination of the demented patient's interests to those of the competent person with labels as self-determination and dignity; see Dresser (1986, 389–390), Dresser (1994, 647–649), Dresser and Robertson (1989, 242–243), Robertson (1991, 9).

Davis, Dena. 2014. Alzheimer's Disease and Pre-Emptive Suicide. *Journal of Medical Ethics* 40: 543–549.
Davis, John. 2002. The Concept of Precedent Autonomy. *Bioethics* 16: 114–133.
Davis, John. 2004. Precedent Autonomy and Subsequent Consent. *Ethical Theory and Moral Practice* 7: 267–291.
Davis, John. 2009. Precedent Autonomy, Advance Directives, and End-of-Life Care. In *Oxford Handbook of Bioethics*, ed. Bonnie Steinbock, 349–374. Oxford: Oxford University Press.
DeGrazia, David. 2005. *Human Identity and Bioethics*. Cambridge: Cambridge University Press.
Dresser, Rebecca. 1984. Bound to Treatment: The Ulysses Contract. *The Hastings Center Report* 14: 13–16.
Dresser, Rebecca. 1986. Life, Death, and Incompetent Patients: Conceptual Infirmities and Hidden Values in the Law. *Arizona Law Review* 28: 373–405.
Dresser, Rebecca. 1994. Missing Persons: Legal Perceptions of Incompetent Patients. *Rutgers Law Review* 46: 609–719.
Dresser, Rebecca. 1995. Dworkin on Dementia: Elegant Theory, Questionable Policy. *The Hastings Center Report* 25: 32–38.
Dresser, Rebecca, and John Robertson. 1989. Quality of Life and Non-Treatment Decisions for Incompetent Patients: A Critique of the Orthodox Approach. *Law, Medicine & Health Care* 17: 234–244.
Dworkin, Ronald. 1994. *Life's Dominion. An Argument about Abortion, Euthanasia, and Individual Freedom*. New York: Vintage.
Elster, Jon. 2010. *Ulysses Unbound. Studies in Rationality, Precommitment, and Constraints*. Cambridge: Cambridge University Press.
Feinberg, Joel. 1986. *Harm to Self (The Moral Limits of the Criminal Law III)*. Oxford: Oxford University Press.
Hallich, Oliver. 2011. Selbstbindungen und medizinischer Paternalismus: Zum normativen Status von "Odysseus-Anweisungen." *Zeitschrift für philosophische Forschung* 65: 151–172.
Hare, Richard. 1952. *The Language of Morals*. Oxford: Oxford University Press.
Hawkins, Jennifer. 2014. Well-Being, Time, and Dementia. *Ethics* 124: 507–542.
Jaworska, Agnieszka. 1999. Respecting the Margins of Agency: Alzheimer's Patients and the Capacity to Value. *Philosophy & Public Affairs* 28: 105–138.
Jennings, Bruce. 2015. Alzheimer's Disease: Quality of Life and the Goals of Care. In *The Routledge Companion to Bioethics*, ed. John Arras, Elizabeth Fenton, and Rebecca Kukla, 437–448. New York, London: Routledge.
King, Nancy. 2007. *Making Sense of Advance Directives*. Washington: Georgetown University Press.
Kleinig, John. 1984. *Paternalism*. New Jersey: Rowman & Allanheld.
McMahan, Jeff. 2002. *The Ethics of Killing. Problems at the Margins of Life*. Oxford: Oxford University Press.
Menzel, Paul, and Bonnie Steinbock. 2013. Advance Directives, Dementia, and Physician-Assisted Death. *Journal of Law, Medicine and Ethics* 41: 484–500.
Menzel, Paul, and Colette Chandler-Cramer. 2014. Advance Directives, Dementia, and Withholding Food and Water by Mouth. *The Hastings Center Report* 44: 23–37.
Moore, George Edward. 1993. *Principia Ethica. Revised Edition. With the Preface to the Second Edition and Other Papers*. Cambridge University Press: Cambridge.
Quante, Michael. 2002. *Personales Leben und menschlicher Tod. Personale Identität als Prinzip der biomedizinischen Ethik*. Suhrkamp: Frankfurt a.M.
Reisberg, Barry, Steven Ferris, Mony de Leon, and Thomas Crook. 1982. The Global Deterioration Scale for Assessment of Primary Degenerative Dementia. *American Journal of Psychiatry* 139: 1136–1139.
Rhoden, Nancy. 1990. The Limits of Legal Objectivity. *North Carolina Law Review* 68: 845–865.
Robertson, John. 1991. Second Thoughts on Living Wills. *The Hastings Center Report* 21: 6–9.
Singer, Peter. 2011. *Practical Ethics*, 3rd ed. Cambridge: Cambridge University Press.

Sumner, Leonard Wayne. 2011. *Assisted Death. A Study in Ethics and Law*. Oxford: Oxford University Press.
von Wright, Georg Henrik. 1963. *The Varieties of Goodness*. Bristol: Thoemmes Press.
World Health Organisation. 2015. 10 Facts on Dementia. https://www.who.int/features/factfiles/dementia/en/. Accessed October 11, 2015.

Oliver Hallich is Professor for Practical Philosophy at the University Duisburg-Essen, Germany. His main research interests include ethics and applied ethics, in particular theories of punishment, philosophy of forgiveness, and bioethics. His recent publications include "Can the Paradox of Forgiveness Be Dissolved?," in: *Ethical Theory and Moral Practice* 16, 2013, 999–1017; and "Selbstbindungen und medizinischer Paternalismus. Zum normativen Status von 'Odysseus-Anweisungen,'" in: *Zeitschrift für Philosophische Forschung* 65, 2011, 151–172.

Chapter 11
Individual and Collective Decision-Making in Palliative and End-of-Life Care

Zoran Todorović and Dragana Protić

Abstract Concepts of individual and collective decision-making strongly influence the bioethical foundations and perspectives of palliative and end-of-life care. Kemp and Rendtorf promoted the integrated approach to basic ethical principles (The Barcelona Declaration 1998) and discussed that "autonomy implies the capacity to make your own decisions about your own life". However, such decisions depend on the interaction/collaboration with others, i.e. their values; the best example is a complex milieu of family communication in the end-of-life care. On the one hand, individual decision-making of palliative care patients could be impaired by their vulnerability, i.e. they are not always able to judge the treatment process and understand/weigh perspectives and possibilities. On the other hand, collective decision-making may give rise to the *bystander effect of collective responsibility*, described by Isabel Baker. The interplay between individual and collective decision-making in palliative and end-of-life care will be discussed thoroughly from the medical practitioner's viewpoint.

Keywords Autonomy · Individual and collective ethics · Decision making · Palliative and end-of-life-care

[1] The Kennedy Institute of Ethics at Georgetown University was founded in 1971 as the first bioethics center. Prominent leaders of the Institute are André Hellegers (first director), Tom Beauchamp and James Childress. They promoted a 'rights-oriented' approach to bioethics, focusing on biomedical issues, and emphasizing *autonomy* as the main bioethical principle. Their approach is deeply rooted in American tradition of personal liberties and privacy rights. In contrast, the 'responsibility-oriented bioethics' affirms *beneficence/non-maleficence* and *justice* as the main principles. Main proponents of such a holistic approach are Van Rensselaer Potter and father of the European bioethics, Fritz Jahr.

Z. Todorović (✉) · D. Protić
School of Medicine, University of Belgrade, Belgrade, Serbia
e-mail: zoran.tdrvc@gmail.com

D. Protić
e-mail: gagamaxi78@gmail.com

M. Kühler and V. L. Mitrović (eds.), *Theories of the Self and Autonomy in Medical Ethics*, The International Library of Bioethics 83,
https://doi.org/10.1007/978-3-030-56703-3_11

11.1 Autonomy

There are three main principles of the American school of bioethics: autonomy, beneficence/non-maleficence and justice. Georgetown School considers autonomy as the most important, i.e. *first among equals* (*primus inter pares*).[1] Being autonomous means that one can make their own decision about their own life. In medicine, i.e. health care environment, autonomy is translated into the principle of informed consent. It is clearly emphasized that autonomy/informed consent is not possible without a competent and well-informed participant, as well as lack of coercion and undue influence. However, Rendtorff conceived autonomy in a different way (Rendtorff 2002). A similar description could be found in Barcelona Declaration of the same author (Rendtorff 1998; Kemp and Rendtorff 2008). According to Kemp and Rentdorff, autonomy means much more than simple informed consent. It concerns the capacity to create ideas and goals for life, the capacity of moral insight and self-regulation, as well as the capacity of political involvement and personal responsibility. However, they emphasized human weakness and dependence on biological, social and material conditions ("autonomy remains merely an ideal"). Hence, everyone seems to be vulnerable in some way. Hidden or not, the concept of vulnerability has been deeply embedded into bioethics from its nascence (Rogers et al. 2012). Fineman even considered it as an ontological condition of our humanity (Fineman 2008). Accordingly, we may consider that autonomy is limited, and humans are more or less vulnerable, which brings us closer to the concepts of autonomy and vulnerability depicted by Kemp and Rendtorff.

In medical terms, autonomy is not *all or nothing*. Rather, it is a continuum ranging from 0 to 100% (no autonomy at all, and full autonomy, respectively). It can be more or less limited for the sake of the patient's health and welfare or for the sake of the whole society (Varelius 2006). Philosophers even argue whether autonomy could be restricted for patients' own good or not (for example, Beauchamp and Childress 2001; Harris 2003). The last revision of CIOMS Ethical Guidelines states that a waiver of informed consent could be granted by a research ethics committee, and trials could be conducted without written approval of participants in a broad range of circumstances from preventive measures to the critically ill. True understanding of information underlying informed consent is of a great importance (Varelius 2012).

Patients' autonomy at the end of life is difficult to maintain, even if consciousness is fully preserved. There are many important issues that have to be taken into account, according to Houska and Loučka (2019). First, patients should make decisions on their treatment and care. Also, they should be actively involved in daily activities, especially to maintain interactions with others. Finally, they should take a proactive role in preparation for dying. All aspects of autonomy at the end of life should be placed in a broader, pluriperspective frame. It is not just a medical issue. There are numerous ethical, legal, social, anthropological, theological and other questions to be resolved. In predominantly Christian societies, there is a different approach to autonomy than in predominantly Islamic countries (Rathor et al. 2016). Rathor et al. emphasized that a unified concept of autonomy is 'not feasible' in the latter case, and it should not be ‘the absolute prerogative of the patient but rather a shared responsibility between the patient, family, and the physician’. Such a viewpoint of Islamic ethicists is closer to the European School of Kemp and Rendtorff than to American principlism.

11.2 Individual and Collective Ethics

At this point, it is necessary to introduce and clearly define the concepts of *individual* and *collective* ethics.[2] These terms were coined by statisticians, not by bioethicists. Pocock and Palmers established and developed these concepts by focusing on clinical studies and the inherent character of ethics from ‘human experiments’ (Pocock 1983; Palmers 1993). On the one hand, we have classical ‘frequency-statistics’ that start from the population as truth, and we consider the individual event in light of probability based on the hypothetical long-term average of the frequencies (for example, if we throw a dice long enough, we expect each number from 1 to 6 to occur 1/6 times). The ethics corresponding to this statistical approach is concentrated on what is best for the patient population or society as a whole. On the other hand, Bayesian statistics,[3] which starts from the conditional probability of an individual event, is considered as a degree of confidence. In other words, in Bayesian statistics, a probability derived from earlier assumptions, experiences and knowledge is adjusted to the information obtained from actual events. This statistical approach is closer to individual ethics, i.e. ethics that defines what is good for a particular patient. Regarding the bioethical issues with the terminally ill, both individual and collective ethics should be taken into account. The former is important in assessing individual patients’ needs, and the latter deals with interests of the whole society.

Recently, the individualistic approach to autonomy in terminal/end-of-life care has been challenged by some authors, and the concept of *relational autonomy* has been extensively discussed (Gómez-Vírseda et al. 2019; Shih et al. 2018). In particular, relational autonomy considers an individual’s relationships with others, i.e. their social milieu, for decision-making. Relational autonomy is closely related to the principles of the European school of bioethics, and it may give rise to shared decision-making. In this regard, it should be emphasized that shared decision-making is usually applied in clinical practice regarding end-of-life care (Elwyn et al. 2012). Such a model may give rise to the *bystander effect*, i.e. responsibility may diffuse across multiple bystanders, diluting the responsibility of each (Thomas et al. 2016). According to Baker, strict and reliable responsibility, accompanied by respective suitable decision-making procedures in hospital protocols, should be defined regarding

[2] It should be noted that the concepts of individual and collective ethics are not thoroughly accepted by some authors (for example, Heilig and Weijer 2005). They rather appeal to a systematic, comprehensive ethical framework.

[3] Thomas Bayes (1702–1761) was an English statistician and philosopher, famous for the theorem named after him (Bayes’ theorem).

decisions in end-of-life care and treatment. The cessation of artificial respiration in ICUs (intensive care units) is a good example of such an issue. First, shared decision-making implies a balance between medical knowledge and attitudes of clinicians and the interests and attitudes of patients and their families. Second, there is a need to combine opinions from different clinicians in order to find the optimal solution. A structured approach to the issue of shared decision-making may involve three main steps: choice talk, option talk, and decision talk (Elwyn et al. 2012).

- *Choice talk* provides the patient with information that reasonable options exist. A stepwise approach consists of careful explanation of the health-related problem, analysis of the available options/solutions and justification of choices. Finally, we should check the patient's reaction and defer closure, if needed.
- *Option talk* consists of several steps as well. First, we should check the patient's knowledge on the health issue (for example, treatment options for cancer). Then, we should list and describe options available with a benefit/risk analysis. Finally, we should provide the patient with decision support and summarize the issue.
- *Decision talk* should focus on preferences, reach the decision and offer a review.

11.3 Individual Versus Shared Decision Making for the Critically Ill from the Perspective of Individual and Collective Ethics

In the last few decades, we have witnessed an increasing complexity of the decision-making process in life-threatening situations in medical practice (Curtis and Vincent 2010). Big data challenges medical providers with an overwhelming amount of information, which are better apprehended than comprehended (Adibuzzaman et al. 2018). There is also a problem of work overload, especially in intensive care units (ICUs). Intensivists could manage such an issue in different ways. They could either make their own decision about the treatment plan for the particular (critically ill) patient or they could share their decision with others. Usually, medical panels with multi-specialty physicians make collective decisions about the most challenging end-of-life issues (for example, treatment cessation in critically ill). However, a more structured approach is usually needed in the vast majority of cases, such as admission to the ICU or discharge from ICU to other hospital wards (Forte et al. 2018). A common problem is overtreatment. Medical doctors tend to overestimate the benefits of therapy and underestimate the harms (Hoffmann and Del Mar 2017).

Medical workers should maintain a subtle balance between their intention to promote beneficence/non-maleficence and to protect patients' autonomy. In other words, the hidden conflict between basic principles of medical ethics should be avoided. In order to manage the constant pressure and to protect patients' autonomy, Western medicine physicians dealing with the critically ill tend to avoid individual decision-making. Rather, they choose shared (collective) responsibility, i.e. to delegate the decision-making to patients and/or surrogates (Forte et al. 2018). Cons of such an approach are clear: the strict promotion of autonomy may deprive patients of the professional guidance that is necessary for reaching the best treatment option. A typical example from medical practice is the conflict between an untreatable patient's wishes (and constant pressure from their relatives), and real (evidence-based medicine) treatment options. Both over- and undertreatment may be wrong in such a case, and a goal should take into account patients' real needs and quality of life.[4]

Foster et al. (2018) offered a comprehensive model of structured, shared decision-making in four steps: (1) ethics of accuracy, focusing on diseases; (2) ethics of comprehension and understanding, focusing on the person; (3) ethics of situational awareness, focusing on providers; and (4) ethics of deliberation, focusing on the patient-provider relationship. The first two steps are conducted by the physician in charge, the third is the task of the multi-specialty panel, and the fourth brings patients and their relatives or surrogates on the scene. Such a model clearly provides a smooth chain of scenarios involving all key players, health care providers and users from a collective responsibility perspective. Both the medical expert approach and wider social perspective are involved with taking into account patients' personal needs and interconnecting both beneficence and autonomy in an essential way. However, such a model of collective decision-making is not applicable in real life, at least in developing countries. Also, justice cannot be respected in a proper way in such a model. In other words, health systems in developing countries may face severe budget limitations. Treatment and care of the terminally ill is usually very expensive (for example, malignant diseases), and cost-effectiveness assessment is usually lacking (McClellan and Tunis 2005). In such cases, physicians and hospital management may be put under pressure by both patients and their relatives, and the health insurance fund. Thus, it is difficult to respect either the patient's autonomy or beneficence, and the ultimate decision is not about what is right or what is best, but what we can actually get done (*the art of the possible*).

We may agree that collective decision-making about the terminally ill is a complex issue. In developing countries, the opinion of health care providers will prevail in the majority of cases, and the autonomy of patients is difficult to respect fully. In such circumstances, the interplay between individual and collective ethics becomes more important from a medical practice viewpoint. Medical expert panels, an important player in shared decision-making regarding the terminally ill, should take into account both approaches. The former, individual one is the primary choice. However, in real life, it is ultimately restricted by the needs of other patients in the hospital sharing the same budget. Medications and medical devices are provided according to the annual projections and budget allocated for those purposes. Moreover, there is a limited number of free beds and increasing number of terminally ill patients. When

[4]Originally, the medical staff usually tends to carry out all possible therapy (according to the Hippocratic Oath) despite the patients' will to cease it and/or the attitudes of his or her relatives to stop with supportive care. Today, the reverse is the case: the hospitals are protected by regulations to unilaterally stop further treatment without fear of legal consequences (e.g. the Advance Directives Act in Texas), while relatives advocate overtreatment.

we put the particular terminally ill patient in a broader social perspective, there is an inevitable shift from the individual to collective ethics from a medical practice viewpoint.

11.4 Palliative and End-of-Life Care

There are many obstacles to the full respect of autonomy of terminally ill patients in developing countries or countries in transition. For example, in Serbia, hospice networks have not been fully developed. Palliative care is a multidisciplinary concept that provides comprehensive assistance to people with incurable diseases, that is, care that increases their quality of life and lessens suffering. It is more successful if the early diagnosis of terminal illnesses is achieved and involves the appropriate pain therapy and management of other issues, such as problems that result from the underlying disease (physical, psychological, social, etc.). Palliative care is provided in hospitals, hospices or the community, and it is conducted by specially-trained, professional health workers or volunteers, both from religious organizations and from particular associations. The basic principles of palliative care are: a team approach, accepting dying as a normal process and integrating the physical, psychological and social dimensions of the human being in providing care (Kennedy 2017; Coyle 2014).

There are two main types of palliative care: general and specialist. On the one hand, general palliative care is carried out holistically, taking into account overall needs rather than the diagnosis of the patient. In doing so, the patient is viewed comprehensively, in general in their environment. This type of palliative care is carried out at the patient's home or in special rooms in nursing homes. On the other hand, specialist palliative care is organized for patients with special needs.

Hospices are of particular importance for end-of-life care. They are designed for the terminally ill (e.g. cancer) whose prognosis is very poor, i.e., the expected survival is short (e.g., up to six months). In addition, there is palliative care for those who are ill with incurable, chronic diseases that require maximum engagement with the community, since often the family deals with a complex and financially exhausting process that sometimes lasts for many years. The difference between hospice and long-term palliative care is not only arbitrary but also has its foundation in the organization of health service. In brief, the National Institute on Aging stated that "… like palliative care, hospice provides comprehensive comfort care as well as support for the family, but, in hospice, attempts to cure the person's illness are stopped." (for details, see the US National Institute on Ageing webpage).

In such circumstances, autonomy is significantly diminished, especially in societies with an inverted population pyramid. The terminally ill are often alone or abandoned by family members, while hospices are not always available. For details about the communication between family members and medical professionals when an individual is incapacitated at the end of life and cannot make a decision for him or herself, see Trees et al. (2017). Resource limitations and limited staff usually preclude full compliance with international standards of end-of-life care, especially in developing countries (Rathor et al. 2016). The liberal concept of autonomy—a cornerstone of the *rights oriented ethics* promoted by the Georgetown School (*André Hellegers, Tom Beauchamp, James Childress*) seems to be a hard-to-reach goal in real life (The Belmont Report 1978).

It should be emphasized that patients are entitled to know the truth, and that palliative doctors and nurses, through successful communication, achieve the basic postulates of this care: an explanation of the meaning of the term 'good death' and the resolution of potential conflicts between the needs of the patients themselves and their relatives. Palliative terminal sedation[5] is not euthanasia for several reasons that will not be specifically elaborated on here (for example, in a palliative sedation, the patient dies of the underlying disease, and not because of the intervention of a physician, and the goal of palliative sedation as a legal procedure is to alleviate the pain, not to end the patient's life).

Besides autonomous decision making, the main ethical issues regarding end-of-life care may involve advance directives, costs involved in the end-of-life care, the physician's role and responsibilities to resolve the issue, and the rationing of care and futile treatment (Karnik and Kanekar 2016). Usually, a multidisciplinary expert panel of medical professionals should reach the conclusion about the termination of futile treatment in the terminally ill. However, such a collective decision-making may suffer from different obstacles, from the conflict of interests to the lack of expertise and vague collective responsibility ("bystander effect", Baker 2011).

In hospitals, intensive care units (ICUs) are of particular importance in providing end-of-life treatment and care. The terminally ill usually die in ICUs after prolonged hospitalization lasting, sometimes, more than a month. Severe bleeding, cardiac or respiratory failure, or sepsis could be successfully resolved, but full recovery is often impossible, and such a patient could not be released from the ICU. Decisions about certain terminal procedures should be reached with informed consent provided by proxies (family members), for example tracheal intubation or terminal sedation (Nava 2012; Warren et al. 1986; Mohanti 2009; Olsen et al. 2010).[6] Usually, free beds are desperately needed in ICUs, and there is a conflict between the family members' pressure to continue with futile treatment of their terminally ill relatives and the need to admit new patients in ICUs.

A careful balance is needed in such circumstances between the physician's opinion (i.e., medical criteria), patient needs, and family decision. The decision on treatment and care of terminally ill patients in ICUs could be even more complex when unexpected (sometimes temporary) recovery occurs. Nevertheless, clinicians ultimately have to rethink whether to continue obviously futile treatment or to reduce it to the necessary symptomatic therapy (e.g. pain management) and let the patient die with dignity (Gurman 2016). Futile treatment could be prevented by drawing up

---

[5]It comprises the continuous use of strong sedatives, e.g. high doses of benzodiazepine (midazolam and others) or opioids in dying patients, aiming at the elimination of suffering and giving the patient dignity in the most difficult moments.

[6]However, in some cases the proxy consent might not be provided due to a lack of family support.

a living will in advance, which is possible in some countries.[7] However, in most circumstances, it is not the case and the decision to stop futile treatment should be made according to the consensus between medical staff and proxies from a medical practice point of view (see above). Such an interplay between different viewpoints (and interests) is tightly related to the concept of autonomy (and other bioethical principles) defined in the Barcelona Declaration (see above). In other words, there is a complex interaction (not just interdisciplinary communication between medical professionals) between health care providers and users aimed at properly sharing the decision on end-of-life treatment.

11.5 Conclusion

End-of-life care deals with vulnerable, dying patients. Successful and ethically appropriate interaction between health care providers and users assumes successful integration of individual and collective ethics in palliative and terminal care that lies primarily in a well-structured communication between providers and service users from a medical practice viewpoint. Many issues should be clarified: specifying the main participants in the dialogue, analysis of the current situation, explanation of the position and unresolved issues of particular decision makers, assessment of the patient's quality of life, protection of autonomy, dignity and integrity of the patient and definition of specific guidance in the given situation.

Literature

Adibuzzaman, M., P. DeLaurentis, J. Hill, and B.D. Benneyworth. 2018. Big Data in Healthcare—The Promises, Challenges and Opportunities from a Research Perspective: A Case Study with a Model Database. *AMIA Annual Symposium Proceedings* 2017: 384–392.

Baker, I. 2011. To What Extent Can We Overcome The "Bystander Effects" of Collective Responsibility in Matters Of Global Injustice? Emergent Australasian Philosophers 4(1).

Beauchamp, T.L., and J.F. Childress. 2001. *Principles of Biomedical Ethics*, 5th ed. New York: Oxford University Press.

Council for International Organizations of Medical Sciences (CIOMS). 2016. *International Ethical Guidelines for Health-related Research Involving Humans* 4th ed. https://cioms.ch/wp-content/uploads/2017/01/WEB-CIOMS-EthicalGuidelines.pdf. Accessed on April 7, 2019.

[7] Advance directives are of particular importance regarding the autonomy of the terminally ill. There are *pros* and *cons* regarding their status. For example, Vogelstein (2016) discusses concerns about the moral authority of such directives. In particular, the advance directive reflects the desires of a fully competent person, while everything could have been changed when terminal illness occurs. In other words, the author promotes the viewpoint that "person's autonomy requires honoring only those desires that are currently attributable to that person". We can easily suppose that the patient may refuse a 'hopeless treatment' as a part of terminal care in his living will, but the relatives may disagree with such advance directives and put pressure on the intensive care medical staff to continue with the futile therapy (for example, antibiotics or cancer chemotherapy).

Coyle, N. 2014. Palliative Care, Hospice Care and Bioethics. *Journal of Hospice and Palliative Nursing* 16 (1): 6–12. https://doi.org/10.1097/NJH.0000000000000032.
Curtis, J.R., and J.L. Vincent. 2010. Ethics and End-of-Life Care for Adults in the Intensive Care Unit. *Lancet* 375 (9749): 1347–1353.
Elwyn, G., D. Frosch, R. Thomson, et al. 2012. Shared Decision Making: A Model for Clinical Practice. *Journal of General Internal Medicine* 27 (10): 1361–1367.
Fineman, M.A. 2008. The Vulnerable Subject: Anchoring Equality in the Human Condition. *Yale Journal of Law and Feminism* 20(1): 1–25. https://digitalcommons.law.yale.edu/cgi/viewcontent.cgi?article=1277&context=yjlf. Accessed on September 6, 2019.
Forte, D.N., F. Kawai, and C. Cohen. 2018. A Bioethical Framework to Guide the Decision-Making Process In The Care of Seriously Ill Patients. *BMC Medical Ethics* 19 (1): 78. https://doi.org/10.1186/s12910-018-0317-y.
Gómez-Vírseda, C., Y. de Maeseneer, and C. Gastmans. 2019. Relational Autonomy: What Does it Mean and How is it Used in End-of-Life Care? A Systematic Review of Argument-Based Ethics Literature. *BMC Medical Ethics* 20 (1): 76.
Gurman, G.M. 2016. Management of the Hopelessly Ill Patient: To Stop or Not to Start? *Romanian Journal of Anaesthesia and Intensive Care* 23 (1): 83–89.
Harris, J. 2003. Consent and End-of-Life Decisions. *Journal of Medical Ethics* 29 (1): 10–15.
Heilig, C.M., and C. Weijer. 2005. A Critical History of Individual and Collective Ethics in the Lineage of Lellouch and Schwartz. *Clinical Trials* 2 (3): 244–253.
Hoffmann, T.C., and C. Del Mar. 2017. Clinicians' Expectations of the Benefits and Harm of Treatments, Screening, and Tests: A Systematic Review. *JAMA Internal Medicine* 177 (3): 407–419. https://doi.org/10.1001/jamainternmed.2016.8254.
Houska, A., and M. Loučka. 2019. Patients' Autonomy at the End-of-Life: A Critical Review. *Journal of Pain and Symptom Management* S0885–3924 (18): 31487–31488. https://doi.org/10.1016/j.jpainsymman.2018.12.339.
Karnik, S., and A. Kanekar. 2016. Ethical Issues Surrounding End-of-Life Care: A Narrative Review. Healthcare (Basel) 4(2): E24. https://doi.org/10.3390/healthcare4020024.
Kemp, P., and J.D. Rendtorff. 2008. The Barcelona Declaration. Synthesis. *Philosophica* 46 (2): 239–251.
Kennedy, C. 2017. Palliative Care. https://www.britannica.com/topic/palliative-care. Accessed on March 27, 2019.
McClellan, M.B., and S.R. Tunis. 2005. Medicare Coverage of ICDs. *New England Journal of Medicine* 352 (3): 222–224.
Mohanti, B.K. 2009. Ethics in Palliative Care. *Indian Journal of Palliative Care* 15 (2): 89–92.
Nava, S. 2012. Non-invasive Ventilation in Terminally Ill Cancer Patients. https://clinicaltrials.gov/ct2/show/NCT00533143. Accessed on March 27, 2019.
Olsen, M.L., K.M. Swetz, and P.S. Mueller. 2010. Ethical Decision Making With End-of-Life Care: Palliative Sedation and Withholding or Withdrawing Life Sustaining Treatments. *Mayo Clinic Proceedings* 85 (10): 949–954.
Palmers, C.R. 1993. Ethics and Statistical Methodology in Clinical Trials. *Journal of Medical Ethics* 19: 219–222.
Pocock, S.J. 1983. *Clinical Trials*, 104–105. New York: Wiley.
Rathor, M.Y., M.S. Azarisman Shah, M.H. Hasmoni. 2016. Is Autonomy a Universal Value of Human Existence? Scope of Autonomy in Medical Practice: A Comparative Study Between Western Medical Ethics And Islamic Medical Ethics. *The International Medical Journal of Malaysia* 15 (1): 81–88.
Rendtorff, J.D. 1998. Basic Principles of Bioethics and Biolaw. Bioethics and Medical Ethics. Twentieth World Congress of Philosophy, Boston, Massachusetts U.S.A. August 10–15, 1998. https://www.bu.edu/wcp/Papers/Bioe/BioeRend.htm. Accessed March 27, 2019.
Rentdorff, J.D. 2002. Basic Ethical Principles In European Bioethics And Biolaw: Autonomy, Dignity, Integrity And Vulnerability–Towards A Foundation of Bioethics And Biolaw. *Medicine, Health Care and Philosophy* 5 (3): 235–244.

Rogers, W., C. Mackenzie, and S. Dodds. 2012. Why Bioethics Needs a Concept of Vulnerability. *International Journal of Feminist Approaches to Bioethics* 5 (2): 11–38.
Shih, P., F. Rapport, A. Hogden, et al. 2018. *Relational Autonomy in Breast Diseases Care: A Qualitative Study of Contextual and Social Conditions of Patients' Capacity for Decision-Making. BMC Health Services Research* 18(1): 818.
The Belmont Report: Ethical Principles and Guidelines for the Protection of Human Subjects of Research. 1978. *The Commission.* Md: Bethesda.
The National Institute on Ageing. 2019. What Are Palliative Care and Hospice Care? https://www.nia.nih.gov/health/what-are-palliative-care-and-hospice-care. Accessed on September 6, 2019.
Thomas, K.A., J. De Freitas, P. DeScioli, and S. Pinker. 2016. Recursive Mentalizing and Common Knowledge in the Bystander Effect. *Journal of Experimental Psychology: General* 145 (5): 621–629.
Trees, A.R., J.E. Ohs, M.C. Murray. 2017. Family Communication about End-of-Life Decisions and the Enactment of the Decision-Maker Role.*Behavioral Sciences* (Basel, Switzerland) 7(2):E36. https://doi.org/10.3390/bs7020036.
Varelius, J. 2006. The Value of Autonomy in Medical Ethics. *Medicine, Health Care and Philosophy* 9 (3): 377–388.
Varelius, J. 2012. On Taylor's justification of Medical Informed Consent. *Bioethics* 26 (4): 207–214.
Vogelstein, E. 2016. Autonomy and the Moral Authority of Advance Directives. *Journal of Medicine and Philosophy* 41 (5): 500–520.
Warren, J.W., J. Sobal, J.H. Tenney, et al. 1986. Informed Consent by Proxy. An Issue in Research with Elderly Patients. The New England. *Journal of Medicine* 315(18): 1124–1128.

Zoran Todorović MD, Ph.D., is Full Professor of Pharmacology, Clinical Pharmacology and Toxicology at the Department of Pharmacology, Clinical Pharmacology and Toxicology, School of Medicine, Faculty of Medicine, University of Belgrade, Serbia. He has studied medicine (1983–89) and obtained his master and Ph.D. at the University of Belgrade, Serbia (FMUB). Specialization and sub-specialization in Clinical Pharmacology, FMUB. Postdoc (1999/2000 and 2001/2002, 12 months) at the William Harvey Research Institute, London, UK (Prof. C. Thiemermann). Visiting Scientist at the Dept. of Hematology, Oncology and Radiation Physics, Skåne University Hospital, Lund, Sweden (2016). Visiting Professor, Icahn School of Medicine at Mount Sinai, NYC (2017). Full Professor of Pharmacology, Clin-ical Pharmacology and Toxicology (since 2008). Former Chairman of Pharmacology, Clinical Pharmacology and Toxicology, FMUB, Serbia (2006–2012). Clinical pharmacologist at the third level hospital. Clinical trial experience: more than 10 phase III/IV studies (role: clinical pharmacologist). More than 300 publications (more than 50 CC/SCI in extenso publications), editor of the international monograph and more than 500 citations in CC/SCI journals. Com-pleted research and education support: National Coordinator, European Pharmacology The-matic Network EPHARNET (EU Socrates), directed by Prof. I. Hughes, UK; the Royal Soci-ety Grant (2001–2002); Ministry of Science of Montenegro, Project Leader; NIH Project 1 R25 TW008171-01A1, Participant. Ongoing research support: Serbian Republic Fund, Co-Investigator, 1990-present; NIH Fogarty Project 2 R25 TW008171-06A1, Participant; COST Action CA19105, MC Substitute. Former President of the Bioethics Society of Serbia (2009–2020). Among other academic and admin-istrative responsibilities, he currently serves as Council of Europe Expert in professional ethics and higher education reform, and member of the Serbian Royal Medical Board. Among his publications are Todorović, Z., Prostran, M.: "Bioethics and pharmacology: 'a voice of one calling in the desert'". In: Schweidler, W. (ed). Bioethics-Medicine-Politics. Sankt Augustin: Academia Verlag, 2012: 133-7; Todorović, Z., Protić, D.: "Bioethical issues in the development of biopharmaceuticals." *Filozofija i društvo* 2012; 23 (4): 49–56; Todorović, Z., Prostran, M., Turza, K. (eds).: *Bioethics and Pharmacology: Ethics in Preclinical and Clinical Drug Development.* Kerala, India: Transworld Research Network 2012 (ISBN: 978-81-7895-579-7); Todorović, Z., Turza, K., Drulović, J. (eds.): *Neuroethics* [In Serbian]. Beograd: Medicinski fakultet Univerziteta u Beogradu, CIBID, 2017.

ISBN: 978-86-7117-521-0; Todorović, Z., Protić, D.: "Patient autonomy and informed consent in critically ill." *Hospital Pharmacology* 2017, 4(1): 463–8.

Dragana Protić MD, Ph.D., is Assistant Professor at the Department of Pharmacology, Clini-cal Pharmacology and Toxicology, School of Medicine, University of Belgrade, Belgrade, Serbia. She has studied medicine (1997–2004), and obtained her Ph.D. in Molecular medicine and specialization in Clinical Pharmacology at the School of Medicine, University of Bel-grade, Serbia. In 2017 and 2018, she successfully accomplished a 1-month clinical shadowing at the Johns Hopkins University School of Medicine (the Kennedy Krieger Institute). Dr. Protic completed the UC Davis MIND Institute\u2019s International Training Program in Neurodevelopmental Disorders (Sacramento, California, US) in 2018. She has a broad background in both experimental and clinical pharmacology, with more than 20 publications, meeting abstracts and several chapters in books in the field of basic and clinical pharmacology, as well as neuroethics. She has broad clinical research expertise as a clinical pharmacologist at the University's tertiary-level hospital, especially her track record of delivering on regulatory pharmacotherapy guidelines, a variety of pharmacovigilance systems, building of intrahospital adverse drug reactions database and pharmacoeconomics experience. Her experience, also involves leadership skills as co-organizer of many CME programs, including Neuroethics mas-ter program and Ph.D. study program in Medical Pharmacology at the School of Medicine, University of Belgrade. She was Secretary for Undergraduate Program, Department of Phar-macology, Clinical Pharmacology and Toxicology. She is member of Steering Committee of the Master Program in Neuroethics, Serbian Medical Association, Serbian Pharmacological Association (member of IUPHAR and EACPT), European Atherosclerosis Society (EAS), KLINIS (Association for the Improvement of Clinical Trials) and Bioethics Society of Serbia. Finally, she also established Fragile X Society-Balkans in Serbia in March 2019. This Society is a non-governmental and non-profit society established with the main objective to organize medical education, research activities, and health promotion in the field of Fragile X. The members of the Society are medical professionals, as well as individuals with Fragile X and their families from Serbia and the whole region. She is the ongoing President of the Society. Currently, her field of interest is neurodevelopmental disorders including autism spectrum disorders and fragile X syndrome. In addition to a lot of publication in this field, her publications include: Todorović, Z., Protić, D.: "Bioethical issues in the development of biopharmaceuticals." *Filozofija i društvo* 2012; 23 (4): 49–56; and Todorović, Z., Protić, D.: "Patient autonomy and informed consent in critically ill." *Hospital Pharmacology* 2017, 4(1): 463–8. She wrote a few scholarly chapters on preclinical neuropharmacology models and neu-roethics in a book entitled Neuroethics, published by University of Belgrade School of Medi-cine, 2017.

Chapter 12
Who Should Take Care of Offenders with Dementia?

Some Thoughts on Fading Selves and the Challenge of Responsibility Interpretations

Annette Dufner

Abstract In this contribution, I investigate the way in which our understanding of a dementia patient's self holds relevance to issues of punishment and responsibility. This topic is motivated by the fact that some countries with particularly large prison populations—such as the United States—are starting to build specialized prison tracts for inmates with dementia. In other countries that do not have such specialized facilities, authorities are trying to find the least badly-equipped facility for such patients, and they are turning to ordinary retirement homes, forensic hospitals as well as ordinary psychiatric and geriatric hospital wards. The problem is expected to become increasingly urgent as the population ages and the number of dementia patients increases. I analyse the way in which justifications of legal (or private) punishment for offenders with dementia can depend on an account of relevant psychological features of the self. I argue that especially retributivist and expressivist justifications of punishment require the offender's ability to comprehend that he or she is being punished for a particular action in the past and that it was him or herself who committed this action. In the second part of my paper, I distinguish between different accounts of responsibility and argue that accounts of relevant features of the self are also needed here to answer the question of whether offenders with dementia are still responsible for past or current inappropriate behavior. In the final part of the paper, I argue that potentially existing private punishment intuitions among caretakers as well as certain puzzles of responsibility interpretation can make it plausible to relieve certain caretakers from primary responsibility for offenders with dementia, especially caretakers who belong to a relevant former victim group of the offender.

A. Dufner (✉)
University of Bonn, Bonn, Germany
e-mail: annette.dufner@uni-bonn.de

M. Kühler and V. L. Mitrović (eds.), *Theories of the Self and Autonomy in Medical Ethics*, The International Library of Bioethics 83,
https://doi.org/10.1007/978-3-030-56703-3_12

12.1 Introduction

What should we do with dementia patients who have committed serious crimes when they were younger and healthy? Should such offenders remain in correctional institutions, or should they live in homes for seniors where caretakers have more experience with serious diseases? Are there any other plausible options? These questions raise a number of important issues about criminal desert and justifications of punishment that are related to the perceived self of these patients. Moreover, as this paper will argue, practical caretaking challenges for criminals with dementia are likely to arise in any setting. Some of these further challenges are closely related to the way in which the past of these patients is likely to influence interpretations of their responsibility for current behavior.

In countries with large prison populations and long sentences—such as the United States—the increasing number of dementia patients is currently prompting the establishment of special prison tracts (e.g. Bernstein 2018). Especially during the later stages of the disease, the inmates of these tracts are obviously very different from what one ordinarily expects of a dangerous criminal, and one may wonder whether a prison is still the appropriate place for them. In other countries, prisoners with dementia might still be the exception rather than the rule. This can also be problematic, because it means that there will most likely not be any specialized facilities for them. In such countries, authorities are reduced to finding the least badly-equipped place for them.

There is now a growing body of literature on dementia in correctional settings. Proposals on how to handle the problem mostly include earlier screening to gain time, creating specialized facilities, or releasing these patients and housing them in other places (e.g. Du Toit et al. 2019). A central question is often whether the person should remain in a part of the correctional system, whether an ordinary prison or a specialized tract. If a patient has a criminal past, one has to ask whether this patient is still deserving of legal punishment for the crime. The first part of this paper will address the way in which desert and justifiable punishment are dependent on our understanding of a dementia patient's self.

Moreover, caretaking for dementia patients requires a substantial amount of interpretive work regarding the intentions, the general mindset, and the human needs of these patients. In any setting, the current behavior of offenders with dementia will sometimes include conduct that seems to imply negative intentions geared directly at caretakers. Caretakers are bound to wonder whether such behavior is merely a symptom of the disease, or a reflection of the "real self" of these patients. The second part of the paper will therefore address the ways in which responsibility ascriptions in such situations depend on the understanding of a dementia patient's self.

Given the uncertainty concerning whether a dementia patient can still be held responsible for inappropriate choices in the present caretaking context, and given the difficulty of discerning any private punishment intentions among caretakers, the final part of this essay will argue that special groups of caretakers should—if possible—have the option to be exempted from being primary caretakers for some offenders with dementia.

12.2 The Search for the Least Badly-Suited Option

The most well-known and most widespread form of dementia is Alzheimer's disease. The symptoms of Alzheimer's include memory loss (especially short-term memory loss), spatial confusion, word-finding problems, and a decline in the ability to keep track of everyday routines. Patients can also display changes in character, moodiness, or an unusual desire to wander around (WHO 2019). In the later stages of the disease, patients become bed-bound. They also tend to lose their ability to speak and to eat on their own (Alzheimer's Association 2020). It is expected that the number of people with a form of dementia will continue to increase (Alzheimer's Association 2020). Accordingly, in the future there will be a larger number of not only (i) prisoners with dementia in the world, but also (ii) defendants at courts who will be able to escape a conviction due to their respective health status. The latter situation can especially occur in cases in which it takes a long time to find the perpetrator, whereby he or she is already rather old by the time of the trial. This can sometimes involve particularly serious crimes such as offenses that do not become time-barred, or crimes that tend to take a long time until all of the facts are confirmed, such as war crimes.

A further situation that holds central concern in this paper is (iii) the onset of frontotemporal dementia, which can cause criminal behavior in formerly law-abiding adults. Criminal behavior is sometimes among the first signs of this form of dementia, when the patient does not yet convey an overall confused or forgetful impression. Disturbingly, these patients sometimes even still possess the ability to verbalize what they did and the fact that it was wrong (Liljegren et al. 2016). This can represent a difficulty for the use of insanity defenses at courts (Berryessa 2016). Furthermore, if these patients are sent to a forensic hospital rather than receiving an ordinary prison sentence, the question arises whether the decision to place them in this part of the system be re-evaluated as the disease progresses, whereby they might have to move again, or whether it can legally be seen as a permanent solution.

It is a serious challenge to ascertain the proper caretaking setting for offenders with dementia. This is especially true for the sentenced offenders in category (i). Due to the probably still widespread lack of ideally-suited options, an ordinary prison will sometimes appear to be a serious default option during the early and mid stages of the disease. However, from a caretaking perspective it has often been observed that ordinary correctional institutions are suboptimal places for dementia patients (e.g. Maschi et al. 2012). Following orders and instructions in a rigid daily routine are key elements of life in prison, and failures to comply with these rigid expectations are easily misinterpreted as intentional disobedience warranting further disciplinary measures. Moreover, the interior design of correctional facilities is usually not adjusted at all to the special needs of dementia patients who need clear signals to find their way around. For example, long rows of identical-looking doors are certainly not ideal in this respect. At the same time, even if there is a proper diagnosis so that correction officers can know about the additional possible source of unadjusted behavior, there are usually no staff with special caretaking expertise or knowledge about the kinds of environmental settings that tend to be beneficial for dementia patients. Depending on the social atmosphere in a prison, further challenges for dementia patients can originate from fellow inmates. Despite some interesting examples to the contrary (e.g. Jaouad 2018), fellow inmates are just as unlikely to possess special training in the treatment of dementia patients as correctional officers. Moreover, stressful conditions in a correctional setting might lead to suboptimal behavior towards fellow inmates who are less able to stand up for themselves.

Given all of these problems, it might appear plausible to consider inmates with dementia as unfit for punishment. In the absence of specialized facilities within (or outside of) correctional facilities, this would speak in favor of releasing them and housing them in ordinary retirement homes where the conditions are geared at the special needs of the aging. In such settings, staff will have proper training and ample experience with medical reasons for unadjusted behavior. Moreover, the interior design of such homes tends to be better suited for people who are confused or tend to wander and get lost.

Nonetheless, this solution unfortunately also presents certain challenges.[1] Although caretakers at retirement homes certainly have more experience with unadjusted behavior of dementia patients than correctional officers, certain problems of behavior interpretation are bound to emerge in any setting. Dementia patients tend to have better and worse days in terms of skills, mood swings and recollections, and it is inappropriate to assume that caretakers would not sometimes wonder whether particular instances of unadjusted behavior are expressive of what the patient was like in the past. Moreover, while correctional officers are trained to cope with aggressive or manipulative behavior and have consented to engage with criminals when starting their positions, staff members in retirement homes might be caught by surprise. Which caretaker in general expects to interact with murderers, sex offenders or war criminals? The attitudes of other retirement home residents and their family members might also be a delicate matter. Which retirement home resident would agree to have such a person as his or her neighbor or even roommate? Moreover, even if the resident does not object, what would the family members say? While there is no general right to know about the criminal past of a neighbor, space in retirement homes is sometimes tighter than in ordinary neighborhoods, whereby in some retirement residences the "neighbor" might sleep in the same room rather than living next door.

[1] A policy-related problem can arise in cases in which offenders do not have any personal funds and the respective jurisdiction does not have a welfare system that will pay for accommodation outside a prison. In such cases, a prison release would require a family member prepared to take care of the patient; otherwise, further imprisonment would seem to be the only ethically defensible option, due to lack of any other possibilities short of dumping an incompetent patient onto the street. The lack of alternatives could also be a terrible burden on potentially available family members.

Given all of these difficulties about the suitability of attendants, the interior design of the building, interpreting unadjusted behavior and the potential fear or inappropriate reactions by neighbors, one might be tempted to suggest a further alternative, such as housing these patients in an "in-between-setting". In the absence of specialized facilities for offenders with dementia, this could—for instance—be forensic hospital wards or other hospital wards with a closed-door policy. It can be argued that such locations would send exactly the right message regarding the special status of offenders with dementia. It might be appropriate to simultaneously view them as both offenders and patients, and forensic hospitals wards in particular would be able to reflect both properties of these people at the same time.

However, it can also be argued that certain problems still prevail. Wards with a closed-door policy are either forensic institutions or they are not. In the former case, the message is that the inmates are still to be viewed as dangerous offenders, which might well be incorrect. Non-forensic wards with a closed-door policy will emphasize the patient's disease more than his or her criminal past, and there could be an issue concerning which characterization is more appropriate here. Moreover, issues remain regarding whether these kinds of facilities are ideally equipped for criminals with dementia. Hospital wards with closed doors are geared at patients with all sorts of different conditions and it is unclear whether the caretakers and the interior design of the ward are truly geared at dementia patients, including dementia patients who are already in the process of dying.

The most suitable non-forensic hospital wards would presumably be psychiatric or geriatric wards. However, arguably such locations would also only be suboptimal solutions. Psychiatric wards are often geared at patients with entirely different conditions, such as depression or schizophrenia. This might leave geriatric wards as the least badly-suited option, although—just like psychiatric wards—these facilities are generally not intended for permanent residence, which is precisely what would be sought in the case of released criminals with dementia.

12.3 Are Offenders with Dementia Still Deserving of Punishment?

Depending on the particular crime that was committed, some readers will not be worried about past offenders ending up in an unsuitable setting such as a prison. Their reaction will be that serious criminals deserve to live a bad life, and that any attempts to find an ideally-suited setting should be a privilege of those who have led a better life. It is therefore worth highlighting that the plausibility of this kind of reaction largely depends on the kind of punishment justification that one takes to be correct. Punishments can be viewed as an expressive symbol (Feinberg 1970), which can either be geared at the public or alternatively the criminal him or herself. In the latter case, it depends on the criminal's ability to understand the meaning of what is being expressed (Duff 2001). This difference largely determines whether punishment can still be justified in case of the demented.[2] Especially if one believes that punishments require a certain level of comprehension and the ability to connect it to past actions to be justifiable, one may ask whether there is still *currently* a being who is deserving of punishment.

In the mid and later stages of the disease, patients will most likely no longer be punishable for behavior that they exhibit in *the present*. They simply can no longer be seen as competent agents who freely and deliberately choose their behavior. However, whether they are still appropriate subjects of punishment for what they did *in the past* is more difficult to determine. First, there is a question of definition. According to general understanding, punishment is a form of hard treatment, and imprisonment normally qualifies as such. However, inmates with dementia might no longer experience their imprisonment as negative. They might not realize where they are and might not miss anything about the forgotten world outside. What they might experience as negative instead are factors contributing to their spatio-temporal confusion, such as long rows of identical-looking doors. Nonetheless, arguably increased confusion due to hallway architecture is not the form of hard treatment that was intended in their original sentence. It may well be an additional, cruel form of punishment for which there is no legal basis.

Second, there is the question of justification. Two of the most well-known strategies of justifying punishments include retributivist and deterrence theories. According to retributivist theories, the offender simply deserves a negative treatment irrespective of whether this will have a positive impact on anyone's future behavior. This view is sometimes supplemented with metaphors such as the idea that the negative treatment is needed to "outbalance the wrong" that has been done, or to "set things straight" again. According to deterrence theories of punishment, a punishment is justified if it can keep either the offender or other members of society from committing similar offenses in the future. This consideration is geared at the future rather than the past.

According to both groups of theories, it is questionable whether offenders with dementia are appropriate subjects for further imprisonment or other forms of punishment. Retributivism works on the assumption that those being punished are still deserving of their punishment. This raises the question of whether desert might not depend on particular mental capacities of the punished, such as comprehension of the fact that they are being punished for a particular action, and an ability to connect this fact to their own actions in the past. In this sense, desert ascriptions might depend on the nature of the self in question (Parfit 1984). Arguably, dementia patients can be seen as fading selves with correspondingly diminishing capacities for desert.

According to deterrence theories, the justification of punishment lies in its prospective force to prevent further crimes in the future. If this expectation is correct, then the greater good of prevention can arguably justify the badness of punishment. However, a deterrence justification might also no longer be plausible in the case of

[2] As I have argued elsewhere, the most convincing and humane justifications of legal punishment require an element of subjective comprehension and self-identifying recollection on behalf of the criminal, which implies that a prison is no longer the suitable place for them (Dufner 2013).

the demented. Punishments can be thought to have a deterrent effect on either the punished individuals themselves or other members of society. A dementia patient does not need any deterrence to be prevented from further aimful criminal actions, and it is similarly questionable whether the thought of further imprisonment during times of dementia will have a preventive force on other members of society.

At most, an expressivist justification of punishment—as it has been suggested by Joel Feinberg—could justify a further punishment of offenders with dementia under certain conditions. According to expressivism, a punishment is a reprobative symbol that expresses the prohibitions and corresponding values of a society. Arguably, an expression needs a receiver in order to work. The receiver could be either the public or the offender him or herself. If the offender is seen as the primary receiver of the expression, the question re-emerges concerning the kinds of mental capacities that need to be present in the offender to understand this message. Again, a certain amount of comprehension of current events and the ability to connect them with a personal action in the past seem to be necessary to speak of a self who can successfully understand a reprobative message about an action that he or she committed in the past.

However, if the general public is thought to be the receiver, an expressivist justification might in fact appear possible. Such a justification—if it is convincing—does not require any particular competencies that a criminal must still have to be a proper subject of punishment. Nonetheless, such an account faces the problem that punishing a human being who no longer comprehends the punishment can be viewed as inhumane. Consequently, the symbolic message of such a punishment would be affected in a paradoxical way. It would seem to express the notion that this society is willing to respond to inhuman behavior with inhuman reactions, which would shed a strange light on the allegedly intended rejection of the criminal's inhumane behavior.

Arguably, one could try to ensure that the particular conditions of punishment would be humane. In a specialized prison tract that is very well equipped to take care of dementia patients, the inmates would not be subjected to inhumane treatment. As a result, the symbolic message would not be affected in the way just described.[3] In this case, at most the problem could emerge that the inmates no longer comprehend the now merely "symbolic" punishment of having to live in a correctional setting. Of course, this issue is ethically less pressing than in case of a punishment that is experienced as hard treatment. Nonetheless, it is a problem that subjecting the uncomprehending to symbolic forms of punishment in order to convey a certain message amounts to treating them as a mere means for this purpose. The suggested form of symbolic punishment—which is not experienced as hard treatment by the inhabitants—should therefore probably have rather tight limits. Perhaps it could remain reserved for cases in which the particular crime holds large-scale significance to the members of an entire nation, such as certain war crimes. In such a case, the name

[3]An attempt to defend such a view has been presented by Hallich, who argues that special prison units where the treatment of the demented would be humane and where they would not be subjected to increased disease-related suffering might send a proper reprobative signal to society at large (Hallich 2018).

and biography of the perpetrator him or herself can have acquired a symbolic meaning that holds relevance to an entire political narrative in addition to the individual victims of the crime.

12.4 The Issue of Responsibility Interpretations

A seemingly common issue for all possible settings concerns behavior interpretation. In both kinds of facilities, the behavior of offenders with dementia can be interpreted as a lack of discipline or being reflective of convictions and habits that they must have held in the past, even when it is in fact caused by their health condition. In the philosophical literature, it has become common to distinguish between at least two forms of responsibility that hold interest here: accountability responsibility and attributability responsibility (Shoemaker 2015; Watson 2004).

Accountability implies that someone is an appropriate target for reactive attitudes such as resentment. It is often said that treating someone as an appropriate target for reactive attitudes only makes sense if he or she is at least in principle able to acknowledge the moral or emotional force of such attitudes. In this sense, it is important that the agent has to be able to say: "I can see and understand what I did to this other person." As Strawson (1962) mentions, a simple scenario in which this would *not* be the case is someone who steps on someone's hand accidentally while trying to help. In such scenarios, we want to say that one should not hold the person accountable for the pain caused. We are inclined to say that there was no willful action in the first place and that only a happening or event occurred. In addition to the capacity to acknowledge what has occurred to others, the relevant behavior must have been expressive of relevant attitudes and commitments of the agent. The agent needs to be able to say: "I can see and understand what these attitudes and commitments *of mine* did to this other person", not just "…what *this foot* did to you."

Attributability requires less than accountability. It does not require that the agent is able to grasp the moral or emotional force of other people's complaints or resentment; rather, it merely requires that a particular attitude or commitment can appropriately be *attributed* to the behavior of this person. Returning to the case of someone who stepped on another person's hand accidentally while trying to help, in addition to the fact that this event cannot be attributed to the person as an *action*, it would be unfair and false to attribute to this person that he or she did not care about the other person. The central concern here is whether the agent's action or attitude discloses underlying evaluative judgments or commitments of this agent.

The question of whether the *current* dementia patient is still responsibility for the *past* offense seems to depend to a particularly significant extent on one's understanding of those features of the self that still need to be present to make such a diachronic ascription plausible. The question of whether the current dementia patient can still be held *accountable* for the past crime seems to depend on a current capacity to acknowledge what has been done to the victim in the past, as well as an ability to identify the relevant action as originating from him or herself, a being that existed then and is now in some significant way still the same. The question of whether the past crime can still be *attributed* to the current dementia patient depends on whether the patient can acknowledge that the crime was expressive of certain attitudes and convictions that were held by a being that existed then and now, as well as whether this being is in a significant way still the same.

In the immediate caretaking context, a further question often emerges: Is this patient still accountable for *current*, possibly inappropriate behavior? Of the two notions, *accountability* seems to require more of an agent. It requires the ability to place oneself into the other person's shoes and see the relationship between one's behavior and the way in which this other person is affected. Acknowledging the perspective of another person regarding one's own behavior seems to require at least a partially shared understanding or interpretation of the relevant action and the situation in which the action took place, which can be difficult in the case of dementia patients. For example, imagine a case in which a patient burdens a caretaker by making noise and getting dressed late at night. A reproach with the content "you scared me and forced me to look after you in the middle of the night" can probably not be acknowledged unless the patient understands that the caretaker has a different understanding of the situation and the applicable normality standards. A dementia patient getting dressed at night is most likely under the impression that there are some good reasons to do so. However, this is not to say that there can never be any accountability during the disease. The loss of capacities in dementia patients tends to be erratic and there may well be temporal pockets of accountability whenever a shared understanding of a situation can permit the acknowledgment of a reproach. However, this capacity is precisely what is on the decline.

When *attributing* responsibility for a more or less *present* action to someone, the central concern is to say that the agent's behavior disclosed his or her relevant evaluative judgments or commitments. This does not include an understanding of the other person's perspective on related behavior. In this sense, it is a less demanding form of responsibility. Nonetheless, constructing a clearly interpretable link between their own behavior and such attributions by others is generally also increasingly challenging for dementia patients. From their erratic behavior, one can often not conclude back to any evaluative judgments or commitments that the patient may hold and that may have led him or her to behave in a certain way. The patient may be living in a bubble that we cannot guess at, and knowledge or understanding of the situation in this bubble might be relevant to construct an intelligible link between behavior and the judgments or commitments that the behavior allegedly conveys. For example, a demented person who does not recognize a caretaker entering the room might assume that this person is her son and then say something like: "you always bug me". Given the referential error that the caretaker may never find out about, he is likely to believe that the snipping remark expressed a particular evaluative judgment about him as the caretaker, although this belief would be false.

This problem corresponds with a problem that relatives often articulate when describing their relationship with a demented relative. They often wonder whether a particular kind of behavior stems from the "real self" of the patient or whether they are confronted with a symptom of the disease. In the aforementioned example, the caretaker is confronted with a symptom of the disease, rather than a judgment that the patient truly holds about him. Interestingly, relatives sometimes revert to the question of what the patient was like in the past to answer the question of whether a particular kind of action is attributable to the patient or whether the behavior is merely a symptom of the disease. For example, if the behavior is something that the patient also used to engage in when he or she was still younger and healthier, the relatives are sometimes more inclined to attribute this to the patient. However, of course this is merely a rather rough attempt to find epistemological guidance. In the case of criminals with dementia, it is bound to lead to disadvantageous conclusions. Healthy young people sometimes change their commitments, and sometimes they even do so more or less out of the blue. Why should this no longer be possible for criminals in the early or mid stages of dementia? In other words, although it might be tempting to solve the puzzle of attributability by appealing to past habits of the patient, this might merely be a rather desperate way of guessing at the truth. As in the case of accountability, this is not to say that nothing can be attributed to these patients anymore. Nonetheless, we have a severe epistemological problem here that might more or less prevent us from making attributions with the correct content.

It is worth highlighting that this issue of behavior interpretation is most pressing during the early and mid stages of the disease. During the later stages of the disease, dementia patients are usually bed-bound, unable to speak, and to eat on their own. At this point in the disease, forms of behavior that can trigger interpretive speculations about intentions are less likely to occur.

12.5 Punishment Intuitions and Responsibility Interpretations in the Caretaking Context

Convictions about desert and punishment among caretakers as well as the difficulty of discerning responsibility for present (and past) behavior of offenders with dementia can create special challenges in the immediate caretaking context. Some caretakers might have the intuition that their patient is still responsible for the crime in some way, and therefore that he or she is also still deserving of further punishment. While the legal question of punishment certainly does not lie within the competency of individual caretakers, they might believe that the patient is deserving of private hard treatment by other human individuals. They might feel a need to state their attitude about this patient's past by not being overly friendly or by being particularly "clear" about certain points that seem to be related to the patient's past. They can also be under the impression that their patient can, to a limited extent, still understand that particular kinds of behavior are undesired, and might be inclined to trigger a "learning process" by subjecting the patient to hard treatment whenever he or she displays the behavior in question. In extreme cases, such a mindset could go unnoticed and lead to a situation that is literally unsafe. Apart from this issue regarding patient needs and safety, it is certainly a strenuous psychological challenge for any caretaker to separate potential attitudes about the crime and the desert of the patient from other attitudes arising during ordinary tasks of daily caretaking.

Futhermore, the puzzles of responsibility—especially for current behavior—can be particularly troublesome when dementia patients behave towards their caretakers in aggressive or other inappropriate ways. In such cases, the question of whether one is confronted with purposeful action or a symptom of the disease can be particularly disturbing and stressful.[4] In the case of former offenders, attempts to find epistemological guidance by considering past behavior is bound to lead negative interpretations. This issue can also result in suboptimal or even problematic situations in the caretaking context. Apart from the potential for negative effects on the patient, in the case of former offenders with dementia the puzzles of responsibility also have to be seen as a special burden on caretakers. Ordinary reactions to the behavior of other human beings include the attempt to ascertain whether there is a message about oneself or something one cares about in what has been said or done. The challenge of caring for a difficult patient not only comprises constantly having to interpret unusual behavior, but also having to block one's ordinary human reactions in many instances. In the case of former offenders with dementia, situations allowing for negative interpretations on the basis of the patient's past can make it particularly difficult to block these personal reactions.

These challenges can make it reasonable to relieve certain members of staff from the task of caring for offenders with dementia. Any caretaker may feel overtaxed by such a situation, and should be taken seriously when voicing concerns. This may be particularly important and plausible for caretakers who belong to a victim group of the offender. For example, it can be a good idea to offer staff members with a certain ethnic background the option of not having to take care of a patient who committed a hate crime against members of the respective ethnicity in the past, especially if some of the patient's current behavior gives rise to relevant puzzles of behavior responsibility. Moreover, relieving a female staff member from care for a former sex offender with female victims can be a good idea.

In the case of caretakers who belong to a victim group of the offender, it can even be seen as an ethical requirement to let them have a say in whether they really want to be responsible for this patient. Making members of staff responsible for the bodily needs of someone who violated a member of a group that they identify with can be seen as placing a rather inappropriate hardship on these persons. They might have a special sense of belonging and solidarity with the respective group—a pro-social attitude that may well be worthy of protection. Arguably, this would make it inappropriate to issue a professional top-down order making them responsible for the patient in question.

However, there are also some problems with the idea of relieving certain staff members of responsibility for offenders with dementia. One problem can be that such an exception might unwittingly signal that the offender is still a danger for

[4]In may be worth pointing out that the opposite situation can also be psychologically challenging. If a former offender wtih dementia is a particularly friendly patient, a self observation by the caretaker such as "I like this former murderer" may also be experienced as troublesome.

members of particular groups. The idea that women should not have to care for a sex offender with dementia because they are women and therefore members of the patient's former victim group can give rise to the assumption that this patient is still physically dangerous, even when this is not the case. The real reason for the exception would not be that an increasingly confused or even bed-bound patient has to be viewed as dangerous (in which case protective measures would have to be taken irrespective of whether the patient has a criminal past), but that there is a general lack of epistemological guidance in interpreting what lies behind more minor forms of everyday behavior. Rather than signaling a potentially questionable judgment according to which this is still a dangerous being, the rationale behind such an exception would be to spare the staff member from doubts about how to read the patient's attitudes and communicative intentions and from having to block ordinary reactions in particularly sensitive areas. It would be important to communicate this very clearly among staff, residents and relatives.

A further problem has to do with identifying those caretakers whose concerns are to be taken most seriously. To stick to the example of a former sex offender with female victims, it will certainly not only be women who will find it particularly challenging to work with such a patient. Clearly, male staff members can also find it overwhelming to have to reign in punishment intuitions and blocking reactions to seemingly responsible and crime-related behavior in such a case. Even though it can probably be argued that such a situation would be even more difficult for a female member of staff, related difficulties among males would have to be taken seriously. Take an instance of potential speculation about attributability responsibility: a male staff member does not have to ask himself whether particular forms of behavior are expressive of misogynist attitudes that have been triggered by his very *own* presence. Obviously, a male caretaker would also be troubled by questionable behavior of this kind and would also have to block ordinary human reactions. However, at least there would be no room for the belief that he has just been insulted based on his sex. Arguably, having to block an ordinary human reaction to such an incident would often be more difficult for a female caretaker. Nontheless, assumptions to the effect that it is a necessary consequence that any female staff member would be more burdened than any male one, should be avoided. This probably speaks against simplified top-down approaches according to which someone makes decisions about responsibilities on the basis of personal assumptions about group sensitivities without previously consulting the relevant staff members.

A final problem worth pointing out concerns the fact that some caretakers might have bad reasons for wanting to be exempt. Some may simply be unwilling to stand up to the special challenge of reininig in their intuitions and interpretations regarding such patients, even though they would be able to do so. Some may even be willing to speak out about their dislike in negotiation situations about work load distribution, which can disadvantage other caretakers who may have more genuine reasons to want to stay away from a former offender. This is a kind of situation that can sometimes require a sensitive, somewhat external perspective in order to be balanced out. Nontheless, it is important to create and maintain a professional atmosphere in which everyone can voice their concerns and in which everyone is taken seriously. Trying to save time by avoiding the complications of a fair process may well lead to particularly undesirable outcomes in the delicate cases of former offenders with dementia.

12.6 Conclusion

As the population ages and the numbers of dementia patients continues to increase, it will become increasingly important to decide how to take care of special populations such as offenders with dementia. In countries in which such patients are the exception rather than the rule and no specialized facilities for them are in place, it is likely that they will not end up in an ideal setting but rather a facility that is the least badly equipped to take care of them. Complementing existing humanitarian proposals, this paper has argued that two concerns should be to ascertain (a) whether our understanding of justified punishment and the remaining capacities of these patients allows for a correctional setting, and (b) among which caretakers their behavior is least likely to lead to particularly painful puzzles of responsibility ascription.

These two concerns will not necessarily point in the same direction. An appraisal of justifications for legal punishment might speak in favor of early release, while the challenges of responsibility interpretation might be more difficult for caretakers in ordinary retirement homes than in specialized facilities within the correctional system. This paper has given no final answer to the problem of where offenders with dementia should live. While there is a growing body of literature about dementia in prisons and other parts of the correctional system, there is still a surprising lack of research on offenders with dementia who live outside of the correctional system. Comparative studies would be desirable before any large scale solutions are implemented in any specific jurisdictions.

Under the assumption that there are options to choose from, and under the assumption that early release is one option, this paper has pointed out some of the more immediate caretaking challenges in dementia care for former offenders. These challenges include the fact that caretakers can have explicit or subliminal views about the criminal desert of such patients, which might have an impact on personal everyday care decisions. The challenges also include the problem of discerning the extent to which a patient is still responsible for current, possibly inappropriate behavior. It is a notorious feature of dementia care that caretakers are prone to turn to past behavior of a patient in order to figure out whether they are confronted with a willful action or with a mere symptom of the disease. In the case of former offenders this can accentuate negative interpretations. Apart from potentially contributing to suboptimal care or even unsafe situations for the patient, these factors have to be seen as a special psychological burden on behalf of caretakers, supporting the idea that some of them should be exempt from having to face up to the task.

References

Alzheimer's Association. 2020. Alzheimer's Disease Facts and Figures. In *Alzheimer's Dementia* vol. 16, 1–391.

Bernstein, Sharon. 2018. California Deals with Dementia Among Aging Inmates. *Reuters*. U.S. News. June 19.

Berryessa, Colleen. 2016. Behavioral and Neural Impairments of Frontotemporal Dementia: Potential Implications for Criminal Law and Sentencing. *International Journal of Law and Psychiatry* 46: 1–6.

Du Toit, Sanetta, Adrienne Withall, Kate O'Loughlin, Nikola Ninaus, Meryl Lovarini, Philip Snoyman, Tony Butler, Katrina Forsyth, and Claire Surr. 2019. Best Care Options for Older Prisoners with Dementia: A scoping Review. *International Psychogeriatrics* 31: 1081–1097.

Duff, Anthony. 2001. *Punishment, Communication, and Community*. New York: Oxford University Press.

Dufner, Annette. 2013. Should the Late Stage Demented be Punished for Past Crimes?" *Criminal Law and Philosophy* 7: 137–150.

Feinberg, Joel. 1970. The Expressive Function of Punishment. In *Doing and Deserving: Essays in the Theory of Responsibility*, ed. by Joel Feinberg, 95–118. Princeton: Princeton University Press.

Hallich, Oliver. 2018. Dementia and Punishment. *Zeitschrift für praktische Philosophie*, 5: 81–106.

Jaouad, Suleika. 2018. The Prisoners Who Care for the Dying and Get Another Chance at Life. *The New York Times*. May 16.

Madeleine, Liljegren, Georges Naasan, Julia Temlett, David C. Perry, Katherine P. Rankin, Jennifer Merrilees, Lea T. Grinberg, William W. Seeley, Elisabet Englund, and Bruce L. Miller. 2016. Criminal Behavior in Frontotemporal Dementia and Alzheimer Disease. *JAMA Neurology* 72: 295–300.

Maschi, Tina, Jung Kwak, Eunjeong Ko, and Mary Morrissey. 2012. Forget me not: Dementia in Prison. *The Gerontologist* 52: 441–451.

Parfit, Derek. 1984. *Reasons and Persons*. Oxford: Oxford University Press.

Peacock, Shelley, Alexandra Hodson, Rhoda MacRae, and Cindy Peternelj-Taylor. 2018. Living with Dementia in Correctional Settings: A Case Report. *Journal of Forensic Nursing* 14: 180–184.

Strawson, Peter. 1962. Freedom and Resentment. *Proceedings of the British Academy* 48: 1–25.

Shoemaker, David. 2015. *Responsibility from the Margins*. Oxford: Oxford University Press.

Watson, Gary. 2004. Two faces of responsibility. In *Agency and Answerability*, ed by Gary Watson, 260–288. Oxford: Oxford University Press.

World Health Organization. 2019. *International Classification of Diseases for Mortality and Morbidity Statistics*, 11th revision, Retrieved from https://icd.who.int/browse11/l-m/en (June 1, 2020).

Annette Dufner is Professor of Philosophy at the Institute of Science and Ethics at the University of Bonn, Germany. She received her Ph.D. in philosophy from the University of Toronto, Canada. After that, she worked for five years as a Research Fellow at the interdisciplinary Centre for Advanced Study in Bioethics at the University of Münster, Germany, and for two years as a visiting professor in the Philosophy and Economics Program at the University of Bayreuth, Germany. She is interested in normative and applied ethics, political and legal philosophy, as well as personal identity theory. Her publications include: "Should the Late Stage Demented be Punished for Past Crimes?," in: *Criminal Law and Philosophy* (2013) 7, 137–150; *The Normative Implications of Personal Identity Theory,* National Library, Toronto 2010 (e-publikation); and "How the Best Life Possible can be Bad: Some Deliberations on Intra-Personal Scope for Distributive Principles," in: *Gehirne und Personen*, edited by Martina Fürst, Wolfgang Gombocz and Christian Hiebaum, Frankfurt: Ontos 2009, 353–362.

Chapter 13
Understanding 'Euthanasia' Across Various Medical Practices

Veselin Mitrović

Abstract End-of-life decisions and assisted suicide are often equated with 'euthanasia.' In everyday parlance of social actors, the term euthanasia is understood rather broadly, even lumped together with other medical procedures. Still, the paper argues that 'intended merciful death,' whether we like the definition or not, ought not to be equated with other practices. Although all of these medical procedures result in the destruction of potential or actual life, the reasons behind such actions could be quite different from empathy or mercy, making the acceptance and advocacy of a problematic definition and understanding of euthanasia the subject of ethical and social debates and analyses. When considering the Universal Declaration on Bioethics and Human Rights, the debate stretches out to include also vulnerable groups in general, which in the contemporary context range from homeless persons and other marginalized groups to embryos created during IVF (in vitro fertilization). The paper presents two case studies, chosen from ten personal stories of former and current IVF procedure patients. In all ten narratives, interlocutors equate abortion with embryo reduction, and both of those with euthanasia. The paper analyzes their perspective to embryos that were not implanted, as well as similarities and differences in their views regarding the activities in the cases of implanted embryos (twin and triplet pregnancies).

13.1 Introduction

End of life decisions and assisted suicide are often equated with euthanasia. In everyday parlance of social actors, the term euthanasia is understood rather broadly, even lumped together with other medical procedures. Still, 'intentional merciful death', whether we like the definition or not, ought not to be equated with other practices. Although all these medical procedures result in the destruction of potential or actual life, the reasons behind such actions could be quite different from empathy

V. Mitrović (✉)
Institute of Social Sciences, University of Belgrade, Kraljice Natalije 45, 11000 Belgrade, Serbia
e-mail: mitrove@gmail.com

M. Kühler and V. L. Mitrović (eds.), *Theories of the Self and Autonomy in Medical Ethics*, The International Library of Bioethics 83,
https://doi.org/10.1007/978-3-030-56703-3_13

or mercy, making consent, advocacy of a problematic definition and understanding euthanasia the subject of ethical and social debates and analyses.

Current debates about this topic are mostly focused on autonomy, generally understood from an individualist perspective.[1] Nevertheless, the approach in this paper demands we also consider scenarios other than only the individualist one.

13.2 Social and Ethical Background

The problem of euthanasia becomes more complicated considering contemporary history and the bitter experiences of various social and political programs, whether executed or on the level of debate and demands made by some social movements.[2] When we consider the Universal Declaration on Bioethics and Human Rights (from the 33rd session of the General Conference of UNESCO 2005), specifically the of human dignity and rights (Article 3),[3] principle of benefit and harm (Article 4)[4] and principle of autonomy (Article 5),[5] the debate becomes meaningful with regard to vulnerable groups and persons without the capacity to consent,[6] which in the novel contemporary context (from medicine to disasters) ranges from the homeless and other marginalized to embryos created during IVF (in vitro fertilization) (Mitrović et al. 2019, 6–9).

With this in mind, and taking a number of indicators (Table 13.1) into consideration, I here examine the role of information held by responsible persons, such as parents in an IVF process, in continuation or abandoning efforts to create life. In what way do they form ideas about their opinions, motives, and reasons? This examination

[1] Tom Beauchamp and James Childress, in *Principles of Biomedical Ethics*, begin their analyses of autonomy "in terms of normal choosers who act (1) intentionally, (2) with understanding, and (3) without controlling influences that determine their action" (Beauchamp and Childress 2013, 104).

This issue is analyzed partially in the context of their three other principles, i.e. nonmaleficence, beneficence, and justice. For Beauchamp and Childress, individual autonomy is preserved wherever it is rooted in some social, cultural or religious belief. However, such social influences are understood as internal characteristics of autonomy. Thus, Beauchamp and Childress essentially still presuppose an individualistic conception of autonomy (Cf. Beauchamp and Childress 2013, 106).

[2] (Dowbiggin 2005).

[3] (a) Human dignity, human rights and fundamental freedoms are to be fully respected. (b) The interests and welfare of the individual should have priority over the sole interest of science or society.

[4] In applying and advancing scientific knowledge, medical practice and associated technologies, direct and indirect benefits to patients, research participants and other affected individuals should be maximized and any possible harm to such on individuals should be minimized.

[5] The autonomy of persons to make decisions, while taking responsibility for those decisions and respecting the autonomy of others, is to be respected. For persons who are not capable of exercising autonomy, special measures are to be taken to protect their rights and interests.

[6] In applying and advancing scientific knowledge, medical practice and associated technologies, human vulnerability should be taken into account. Individuals and groups of special vulnerability should be protected and the personal integrity of such individuals respected.

Table 13.1 Tabular representations of results

Participant	Euthanasia is the same as…	Self/autonomy in IVF (Participant view)	Reasons for abortion	Reasons for embryo reduction	Scope of euthanasia	Prevention of euthanasia	Eugenics and euthanasia
P3	Abortion, embryo reduction	Individual stic perspective	Malformation economic	Care	Narrow	Positive	Consistent
P8	Abortion, embryo reduction	Individual stic perspective	Malformation right moment/partner	Dear self	Wide	Negative	Consistent

also helps us to better understand how patients undergoing IVF themselves understand the procedures in which potential life is either created or destroyed. We must not forget that creation of embryos in vitro only carries the potential of life. Therefore, if one such embryo is offered the possibility to live (meaning that it will become a person), then we must take the same care in the course of deciding to destroy this potential life. In other words, the question becomes, why do parents undergoing IVF give certain embryos a greater chance to become life than others? This is the crucial question for understanding euthanasia through other medical practices, but also for avoiding justifying eugenic choices (Mitrović 2015).

Contemporary technology allows us early insight into some embryo defects, thus avoiding certain serious illnesses.[7] However, could the same technology be applied not merely to avoid defects, but to select for desirable ones, thus conducting a morally questionable selection among vulnerable embryo and potential human beings?

As we will see in this study, this approach to creation of offspring allows for the destruction of life for the sake of the most selfish of goals (dear self).

With these questions in mind, I present here a portion of a broader study conducted with IVF patients. More specifically, this paper presents the positions, conceptions, motives and experiences of embryo reductions and pregnancy termination (almost always equated with euthanasia in the case of certain fetus anomalies).[8]

Still, some of the participants in our study undertook these practices (which they themselves equated with euthanasia) for other reasons. Particularly interesting are the cases of twin or triplet pregnancies, which generated entirely diverse explanations for terminating embryo life created through IVF and implanted into the mother's body. Further, I examined other potentials that could influence such a decision or play a role in maintaining or changing this equating of different practices.

In this article, I will focus on the stories of Participants 3 (P3) and 8 (P8), chosen from ten personal stories of former and current IVF procedure patients. I analyze their perspective on eggs and embryos that were not implanted, as well as similarities and differences in their views regarding the activities in the cases of implanted embryos (triplet pregnancies). Patient P3 declared herself as religious, belonging to the Serbian Orthodox Church, even though she added that she did not go often to church. At the time of research, she was an unemployed high school graduate. Participant P8 was atheist, highly educated and employed full time. In the course of IVF, both were implanted with three early stage embryos. P3 gave birth to healthy twins, while P8 gave birth to a single healthy child. In addition to a basic goal, these stories reveal how verbalizing a narrative arrives at the authenticity, motives, and reasons that contributed to making a given decision. In other words, the interlocutor is never merely reporting on a series of data or information, but is always the narrator of a

[7]Mark S. Frankel. 2003. "Inheritable Genetic Modification and a New Brave World: Did Huxley have it wrong?" in The Hasting Center Report. 33 (2): 31.

[8]For detailed insight into the questions used for this research, see appendix, in Mitrović 2016.

multifaceted story (Holstein 1995). And since the narrative is generated in the very course of conversation with the researcher, the conversation can as such be the result of representation and analysis of empirical material (Hyden 2008, 50).

13.3 Self and Autonomy in Procreative Decision Making

In addition to the principle of nonmalefience and the principle of beneficence, respecting patient autonomy represents a significant role in making medical decisions, in particular when it comes to crucial life decisions (Beauchamp and Childress 2013, 104–149). Respecting Autonomy, that is, its understanding, as we shall see, has a decisive role in the decision and acts that would save, bring good, or destroy current or potential life. Such a scenario is especially important in the case of those vulnerable, without the power of choice or consent (Mehlman 2009). The question, from the perspective of those making the decisions, is how is one's own image and one's idea of autonomy formed, what is the partner's role in it, as well as of mutual trust within a family? Is autonomy something that can be lost or acquired in the course of one medical procedure, only to ultimately result in a negative result for some, instead of being of benefit to all involved—starting with the respect of autonomy in the number of embryos and the right to further information and procedure consent, to selfish embryo reduction.

The first challenge to formulating limits to autonomy, trust, and making joint decisions regarding creation (and later destruction) of life can be found in the portion of the conversation about the initiative and deciding on IVF. It is characteristic for men to think it entirely natural that the women take the lead as well as have the last word on the matter, while for their part, a common position was the slogan "your body—your decision." This approach to the issue of infertility, as a challenge to young married couples, at first seems like something that could significantly influence later decisions, meaning the joint demand for preserving all potential lives that could be preserved. It can also be understood as a matter of trust in the partner's decision, the way of formulation of the self in interaction with the partner. However, such a socially-conditioned approach to personal choice (Kühler and Jelinek 2013: xii–xvi) can also carry another, more latent significance and avoidance of responsibility in a joint decision (Beauchamp and Childress 107), and consequently have the opposite effect—mistrust. With this in mind, the rest of this text should be used to examine whether autonomy constituted thus carries the potential of endangering all potential lives, and whether it also plays any role in conceiving possible rationalization of an entirely different practice (Childress 1997), unconnected with the couple's reproductive familial challenges—euthanasia.

In all cases, participants or their partners answered the question of their partner's involvement in a similar way.[9] Aside from the answer to the question who was more responsible for the initiation of the IVF process, participant P3 also gives the influence of her surroundings:

> It was my initiative.... As opposed to the pressure from my family, that I should give up on this impossible mission, to withdraw from all this, that if I want children I should be making them with someone who has a chance, I even had my own mother telling me for four years, not to do this to myself, not to go through these hormone stimulation, not to put myself at risk for uterine or ovarian cancer, but that I should find another partner for fertilization. [laughs] However [laughs], erm, of course, I understood all this, I mean, she's a mother, caring for her child. I am after all her kid, unfortunately, she is less interested in my children. Simply, people become selfish. They want their kid to be healthy, but what that child wants and what makes it happy, that is already a luxury in Serbia.

Describing the role of the partner in the IVF decision, that is, whether that role was exclusively made based on the desire for children, or whether it was an issue of the survival of the relationship, she vacillates. There is a contradiction, that is, the verbalization of the problem. In other words, it is in the course of the conversation that she becomes aware of certain motives and reasons that influenced the decision:

> [...] Hmm, no, not the relationship's survival, we both wanted children, even when we discovered that it is a matter of infertility, this really frightened us, and the road to IVF is really daunting. Because of the decisions, the paperwork of our health system, because it tests the relationship, because there is no guarantee that you will ever have children, although you invest a lot, and to be honest, I was quite near to leaving the relationship, precisely because of all the pressure, it's all on your back, you know. Simply, it changes everything, it changes your relationship with your partner. The partner who is sterile withdraws, erm, even if we had good relations before that...Everything changes, everything changes. It's shaken, and that's that...I really do not know how we made it. I am sorry that we didn't seek psychological help because that would have made it easier on both of us. Someone to talk to, but there was no such thing in Serbia. Now, there is a psych consult clinic on Svetogorska Street, I know all that, I even know some people who work there, they spoke to me about my own IVF, anyway, but in the clinics there is no psych help. Abroad, every clinic has it, you don't have to, but they strongly recommend it. I am sorry that for my first IVF, even this second one, which was conducted by the most "European" standards, because it was done in a different city abroad, I am sorry that I did not seek psychological help on time, but only after giving birth. Now I go to all the workshops, consults, to a therapist. Yes...it is very difficult for the relationship to survive (P3).

Telling us when she decided to undergo IVF, participant P8 says:

> [...] When the doctors told me that we would never be able to have children naturally. So, after the information from the doctor. I was told that with these results, the two of you will never conceive naturally. Simply, my husband did not have good results (P8).

To my question whether in addition to this medical information, their decision was also motivated by preserving the relationship, she answers: "No, no. I am sure that it was not so important to us" (P8)!

---

[9]For example, a characteristic response of the partner was: "She lost patience, so it was her initiative, I had nothing against it, it was up to her. In May, we approached the clinic, the insemination was in July..." (P7).

Moving to the description of the partner's role in deciding on IVF, she told that: "He said, this is your matter, your body and you have to decide! I am for it, but your decision is crucial" (P8).

13.4 Three Versions of Abortion

Of the ten personal stories from patients taking the path of getting around sterility by way of ART (*Artificial Reproductive Technologies*), nine supported abortion, that is, considered it legitimate in similar, but also in different, circumstances. I found complete support for terminating pregnancy in cases when participants were faced with reliable information pertaining to irregular development of the fetus. In these cases, abortion was considered a kind of euthanasia, regardless of whether the participants were religious, agnostic or atheist.

In the course of their pregnancies, some of the patients were faced with information that the fetus is not developing normally (based on one of the tests); still, they did not reach for abortion right away, but consulted other experts who explained that the fetus development cannot be ascertained with absolute certainty through the obtained test results. The participants were asked, "if a routine test established that a potential fetus had certain anomalies, to what extent should the doctor go in informing the future parents about the quality of life of such a newborn?"

Participant P3:

> Well, so, I did the whole embryo reduction, after which I did the amniocentesis. Which means the genetic examination of the baby. I looked at whether I am at risk after the second penetration of the uterus, and I did the amniocentesis and one child has Down syndrome, then you do some five genetic tests, what happens with a twin pregnancy and after that test if one does have Down syndrome and it is already the fifth month, while the other does not – what happens then? The consilium just yelled at me and said that I best think of nice things and get out. In other words, they did not have an answer but were quite eager to offer the second amniocentesis. Which I accepted, but what happens if one child has Down syndrome, I cannot give up on it for the sake of the other, I give birth to two kids whatever they had. When it comes to twin pregnancy, I think it is not ok to do amniocentesis since the likelihood is small for both fetuses be damaged. Luckily, our results were great and this is a great risk for twin pregnancy (P3).

In order to obtain an even clearer answer, the participant was asked the following question: "What would you have done had one child turned out to have Down syndrome, would you save one and not the other?"

> There is an earlier and later amniocentesis, and I did the later. Had I found out that one had [Down syndrome], well I certainly would not kill it in the fifth or sixth month, I don't know, I would leave it. If the earlier were done, and that were the result, I would remove it. Simply, this is the answer why women are odd when pregnant. Really, you think too much or not at all. But I did the amniocentesis and all that, but to no avail (P3).

Even though participant P3 had no experience with abortion (she had embryo reduction), her position regarding that practice was that:

> My position is that I support it absolutely if people are not ready, either psychologically or existentially to bring up a child. But I am also against it, I mean, since there are forms of protection. Since I think it's a trauma for the person, thinking about abortion, that it is some kind of mere extraction. I think that this is a trauma, torturing both the woman and the man, that there is no reason for abortion, maybe in the case of rape or something, but not just for the heck of it, no (P3).

Six of eight patients (two conversations were held with men) had personal experience with abortion, pregnancy termination, or embryo reduction. In these cases, in addition to malformation, reasons given ranged from unplanned pregnancy (not the right moment or not the right partner) to financial unpreparedness for raising children.

Participant P8 was asked the same question of what would happen if routine control showed that the potential fetus has certain anomalies and the what extent, in the participant's opinion, should the doctor go in informing the future parents about the quality of life of such a newborn.

> [...] Well, the doctor has every right to inform me. Without telling me what to do (P8).

I continued with the follow up question whether she would terminate the pregnancy if a test result indicated an increased risk of anomaly in the newborn? She answered in the affirmative.

As to her direct experience with abortion, and reasons for terminating pregnancy, given the efforts to treat sterility, the participant responded:

> I had an abortion, erm, some four years before IVF. It was a pregnancy termination. What is my position? Positive, since I can choose who will be the father of my child. That's what I think [...] [laughter] (P8).

Another follow up question was whether she tied this decision exclusively to the autonomy of the mother, that is, the woman? She answered in the affirmative. In elaborating her reasons for terminating pregnancy, she answered briefly and symbolically:

> A dash outside of marriage [laughter], a dash outside of marriage [laughter] (P8).

The equating of abortion with euthanasia acquires significance when we compare the experiences of patients who despite such a view of abortion, conducted such destruction of life for other reasons. For now, this contingency in conceiving of abortion carries the potential of an ethical transition of justifying a means for achieving entirely disparate goals. In the following portion of the research, I will examine whether this potential can be achieved through a combination of experiences of different individuals in the same situation.

13.5 Embryo Reduction: Care, and Selfishness

There are clear differences between the statements given by P3 and P8 regarding embryo reduction, that is, the limits of decisions and motives for it. In the case of participant P3, we can assume there was care for the remaining implanted embryos. The care for their health and survival had a crucial influence on the destruction of one of the three implanted embryos. Still, the answer also features an economic factor connected to the possibility of child-raising if they are born with defects. This background factor gives another reason for justification of embryo reduction in the case of triplets.

> [...] Well, I did have embryo reduction. Three embryos were returned to me, even though I asked what if they are triplets, but the doctors said that they would turn out twins, it's only 5% likely that they would triplets. But I asked what if it turns out I am in those 5%? They said I would figure it out when it happened, now it is important to get pregnant. I understand their position. They asked if I wanted twins and I said yes. They gave me back three embryos and they turned into triplets. So I had a difficult decision to make. Do I remove one, as one doctor abroad who does IVF recommended, just like local doctors? Then I risk all three babies being born prematurely, that is, that I have them, erm..., or rather not have any or have them, but one with some defects, etc., due to the lack of oxygen and all that. A twin pregnancy is also quite risky. So I thought whether I should leave all three, that I do not have the money, I have no inheritance, I live in Serbia where child protection and health care is not a high priority. So, I thought how I would help these children if they have some difficulty, meaning developmental. Since they were three in my belly. Then I decided that since I am not financially strong enough, either considering my family or individually, to remove one anyway. It was a very difficult decision. But simply, I consoled myself and said ok, you underwent IVF and you know the potential risks as well as having two kids, and giving them the chance to be born healthy and strong, I had to remove one (P3).

Since she had not had direct experience with abortion, but had had experience with embryo reduction, I returned to the question of pregnancy termination in case routine control indicated increased risk of anomalies in the newborn. Further, I repeated the question of the doctor's influence on the decision. In other words, how much does the doctor, with their suggestion, influence the mother in her decision-making in these cases?

> [...] Well, I felt they had influence regarding embryo reduction. I am thankful to our doctors, they told me, that is, one really good woman doctor told me, I saw triplets, they look fine, I don't know if yours will be like that, but I saw these and others who had something wrong with them. Think about what you are ready to do and what you... It's simply your decision whether to take the risk or not. Which is the same as what they said abroad. They told me that chances were 50% that you could give birth to all three babies dead if you leave them. If you undergo embryo reduction, you have 80% chance of having two healthy children. When someone gives you 80% chance and those other 50%, it really makes it easier. But not our doctors. They stay out of it. They just say that anything is possible, and it's up to you to decide. Abroad they gave advice.
>
> [...] Abroad they told me that there is 80% chance to give birth to two healthy children. They know what the embryos are like, they even thought that I do not need to do the amniocentesis, since the embryologists knew in advance what they had done. I mean, experience is very important (P3).

In the case of participant P8, after the IVF treatment, one child was born, even though she had been implanted with three embryos. When I asked whether she had undergone embryo reduction and whether she saw this as termination of pregnancy or abortion (previously, termination of pregnancy of a fetus with an anomaly was equated with euthanasia). The answer was that:

> Yes, I did. Yes, I do see it as termination, and it was done in another local private clinic. One embryo disappeared on its own, while one was removed by me (P8).

When I asked why she made that decision and whether she was following medical advice, she said:

> No. I made the decision on my own because I know myself. After a few weeks, I was thinking about the twins, and I knew that I would not be able to do this, how am I supposed to support even one, no less two. I know I'll need help that I don't have. For me, the way things are, one child is what I can have (P8).

When I asked whether she considered early birthing due to the twins, she answered: "Yes, I did, and it turned out that had I not done that, I would not have had any children! But I could not have known that. I was going on knowing myself" (P8).

The two cases are also textbook examples of the discrepancy between ought and can. For P3, the principle of "ought implies can,"[10] means that all potentially salvageable lives ought to be preserved. But this principle was broken due to medical realities in which not all embryos could be preserved. On the other hand, for P8, this principle plays no role: potential life that might have otherwise developed was destroyed for selfish reasons.

Accordingly, we could speak about normative and social differences between regret or mourning and blameworthiness or guilt retrospectively (Kühler 2013, 207).

13.6 Scope of 'Euthanistic' Action

The scope of 'euthanistic' action is an imaginary scope of autonomous decisions that refer to the preservation or destruction of potential life created in the course of IVF. To obtain answers regarding this scope of action, I used a set of questions that refer to the *pro vivo* position about the remaining reproductive material after the IVF. Such questions helped us understand the decision about embryo reduction in these cases, since the initial relation of partner trust and the constructed self as well as the degree of autonomy in the initial decision would appear to be reflected onto all other decisions, but also onto a lack of moral support in cases of ambivalence regarding preservation of potential life.

In P3's statements, we see a *pro vivo* position towards the remaining cells, but the reasons are not for the preservation of this potential, but rather of economic nature.

Answering several connected questions about whether she would preserve all the embryos obtained via IVF, participant P3 said that the clinic staff actually preserved everything even though some of the material divided "slowly." Still, she got no guarantee from the embryologist that all five embryos are developing as they should.

> [...] when I arrived, three were alive, they let them all divide and take photos of them constantly, they watch them divide and follow the fragmentation, etc. The other two simply

[10] For more about principle "ought implies can," see in Kühler 2017.

> stopped dividing, they died. I don't know how to say it, but these three were excellent, although I have an image of them and one really has quite a few fragmentations. So, we do not know whether this is one of these two twins today [laughs]. You know what I mean? Each embryo has to be frozen. If you ask me, no doctor or embryologist will tell you this will or won't be a baby. Even those "lame"[11] ones could well become babies (P3).

Answering what she did with the other, early embryos after IVF, participant P8 said:

> [...] How many there were, I don't know, but three were good, and they were returned. No one told me, nor did the discharge papers say anything like that. I didn't insist, since that clinic does not freeze, so I didn't ask. I agree that they should have asked what we decided to do with our reproductive material, but at that moment, erm...how can I put this, if you decide to trust the doctor and everything is in his hands and the lab, then you simply do not ask. I had a different dilemma, which was that during hormonal hyper-stimulation, over thirty eggs were taken out. I wanted to know, where are all my eggs? I didn't freeze them, I didn't do anything with them, so where are they, what happened to them? If there is no possibility of returning them without being frozen, what's the third option? [laughs].

13.7 Prevention of Euthanasia

Opposing the potential of euthanistic action, I proposed the possibility of preventing what the participants took to be euthanasia, by destroying the fetuses and early-created embryos. This potential was measured by way of questions about potential donation of obtained embryos, eggs or semen for scientific research intended to prevent the very defects that lead to the euthanistic behavior in participants.

Participant P3 answers the question affirmatively:

> Research—always. Right away. Because it was thanks to research that I have these two children. That is, thanks to science, let's be honest. However, when it comes to donation, should they be thrown away or given to someone else to have a child...Well, I think I would give them away, I think that this is best (P3).

In the statement of P8, the question of donation of embryos for the purpose of research provokes an ambivalence caused by mistrust in scientific work, and it results in a negative answer.

> [...] possibly for the sake of research. Although I have doubts. It would depend how much information I would have what's being done with the embryo, which I do not believe I would get here in Serbia. So, I would say no. Better to destroy it right away than have it developed and then have something done to it. I would most like to destroy my embryonic cells (P8)!

[11]The participant used the Serbian word *kljakavo*, an extremely derogatory version of lame or damaged. The patient is using the word ironically, because during the IVF procedure, at a private clinic in Serbia, one of the nurses used it to describe the embryos being implanted in her uterus, letting her know that the odds of a baby are slim.

13.8 Eugenics and Euthanasia

I thought it necessary to also examine the autonomy of participants regarding the possibility of creation of a perfectly healthy and, if possible, enhanced offspring (Harris 2010). The participants who developed their identities partly on the idea of reproductive difficulties, achieving parenthood, but also the possibility of justifying the destruction of potential life (which they considered euthanasia), were faced with the question of limit to which they would go in choosing characteristics and capacities of their future child. For example, avoiding chronic illnesses, choosing the sex, eye color, height, cognitive abilities, sexual orientation, etc.

A particularly dangerous potential in both cases comes from the subjective perspective and rationalizing euthanasia in the form of abortion of the fetus due to a defect, and the demand for a healthy and enhanced offspring, whether by way of reduction of the initially weak and asymmetrical embryos or by detection of illness.

To all this, from participant P3, I received the following answer:

> [...] For illnesses, that can be done at the clinic where I got my IVF. On the embryo. Which is expensive. I would absolutely do it if I had the money, and if I had...erm...persons in my family who suffered from genetic disorders, etc. Absolutely yes, I would not give birth to such a child (P3).

Answering the follow up question, whether she would go further, and choose the sex, eye color, height, cognitive abilities, sexual orientation, etc., she said:

> [...] The rest...you know, people often tell me, you have two sons, it will be tough with two daughters in law. To which I answer, what if there is a son in law! But I really, not only do I not prohibit it, but just as I would be happy to have two daughters in law, I would be happy to have, not one, but two sons in law. I would not do any of those... I would not choose those things. None of that (P3).

It is thus clear that participant P3 would not choose the potential sexual orientation of her offspring. Later on, in the conversation, she adds:

> [...] Maybe in twenty years, maybe tall people, just as pretty people now, will have it easier. Erm...yeah...Anything would help them (P3).

> [...] So, I would, if Serbia became a great place to live, erm, you know, I look at it from a global perspective. What if they live in a country where tall people are not prevalent? It would simply have to be global. That is, I would take that step based on world trends in twenty years. Do we know or not? Perhaps there will only be tall people or only short people, and in that case I would of course choose to have kids accordingly, I would think about that, yes (P3).

Participant P8 stated that she was not aware of the possibility of controlling obtained embryos, although part of her story referred to an event in the lab when she was able to follow the development of her early embryos.

> [...] Yes, but what does that mean? Does he have to divide in the course 1, 2, 3 days or so? I have no idea whether that is a good thing, looking through a microscope. I have no idea how it's verified. I think that it is not even necessary to be explained how it's verified since he is the doctor and he knows the procedure... and the third day was the return of the embryos (P8).

Later in the conversation about potential avoidance of chronic illnesses, choice of other characteristics, abilities and capacities in her future offspring, she says:

> [...] Avoiding illness, yes, certainly. Sex and the rest, no (P8).

Asked whether she would genetically predetermine or enhance other abilities of her future child, such as intelligence, she answers in the affirmative:

> [...] That I would, why more stupid when it could be smarter (P8).

Asked whether she wished for a more empathic or more rational child, she said that she would not change anything in that case, and just like the other participant, she found that it is best that the child be empathetic and altruistic. The two cases held a common position that it is more ethical to turn to the study and manipulation of bodily, rather than embryonic cells. While P3 was more prepared to donate her genetic material in accordance with the altruistic drive, P8 was not ready to take that step.

13.9 Conclusion

Although both participants were entirely convinced that during the course of the IVF, they followed their own independent decision, during the conversation and verbalization of certain challenges, they arrived at the conclusion that their decisions were influenced by several different factors, such as the environment, their partner, doctors, etc. Often, the preservation of the relationship, as a latent social pressure, played a role in initiating this procedure and all its consequences. In addition, the neutrality of the partner regarding certain decisions gives an ostensible picture of autonomy. This form of decision-making could later be connected to the idea of acceptance of one's powerlessness to raise multiple children or relinquishing to fate the remaining eggs and early embryos. Further, there are the suggestions by the doctors, grounded in test results, that range from hyper-rational prognosis about the future health of the child, to awarding equal odds to the birth of healthy triplets as giving birth to triplets with defects, unless embryo reduction is conducted.

In the course of uttering and constructing the story in conversation with the researcher, both women often used hesitation markers such as "erm," or laughter in those places where, according to them, their choice does not respect the autonomy or integrity of their partner, or else is not socially acceptable.

Almost all patients considered abortion justified as a means of family planning, either in choosing the timing of the pregnancy or as termination in the case of fetal anomaly. All participants equated abortion and embryo reduction. Participant P8 had an abortion before the IVF process and embryo reduction during it. Participant P3 had embryo reduction during IVF. All the women in the survey encountered one or more of these experiences prior to or during the IVF process. They justified abortion and embryo reduction even as a type of euthanasia.

Participant P3 took the embryo reduction in a difficult way, even though it was recommended by doctors of a clinic abroad (where the IVF was conducted). Their explanation was that there was a high probability of defect or death of one or more of the embryos is she carries the triplets to term.

Several years on, she still feels the psychological effects. There is a heightened sensitivity and care for life and health of the twins, the result of the justification of this embryo reduction. In that sense, she remains of the belief that had she had the money, she would have preserved all the created early embryos, regardless of their initial irregularity in the first hours of cell division.

By contrast, participant P8 chose embryo reduction because she thought that she could not raise two children. This decision, in addition to a selfish motive, incorporates an irrationality tied to the risks of medical procedure entered and a decision to destroy a potential life created during this procedure.

The reasons for destruction or leaving to fate the unused genetic material are not only personal, but often socially determined, starting with economic to ethically questionable practices in the clinics. In the case of participant P3, the reason was exclusively economic in nature, while for participant P8, the reasons were relinquishing care and decision-making to the doctors.

In different circumstances, participant P3 would be prepared to donate her eggs or early embryos for research purposes. After verbalizing the problem, participant P8 hesitated and initial agreement ended up in a decision not to agree to a donation, broadening the scope of what was understood by her to be euthanasia.

Considering all the similarities and differences followed through these indicators, the conclusion I can offer is that the greatest danger lies in social and ethical justification of a practice that has nothing to do with infertility issues these couples face. Both participants were guided by the notion that there is not much difference between "advantageous" principles of Darwinian evolution and "Enhancement Evolution" as some scientists understand it (Harris 2010, 11) . Accordingly, they were ready to use biomedicine to enhance their offspring, to a certain degree and in accordance with the global social demands and trends. But we must not forget that "Darwinian evolution" is neither moral nor smart, but powerful and opportunistic. With that in mind, the contingency of eugenics and euthanasia gives rise to a dangerous and powerful social potential, which coupled with a loss of a traditional approach to parenting (Almond 2006) and general care for the weak, can produce unforeseeable consequences in modern society. Hope remains in the knowledge that these practices can be prevented by an active engagement in solving the problems faced. Interest and insistence regarding the genetic material left over in the course of IVF, as well as the solidary and empathic decisions (as in the case of P3), could help in medical research and treatment for infertility and hereditary illness.

References

Almond, Brenda. 2006. *The Fragmenting Family*. Oxford: Oxford University Press.

Beauchamp, Tom L., and James F. Childress. 2013. *Principles of Biomedical Ethics*. New York: Oxford University Press.

Childress, James F. 1997. *Practical Reasoning in Bioethics*. Bloomington: Indiana University Press.

Dowbiggin, Ian Robert. 2005. A Concise History of Euthanasia. Life Death God and Medicine Lanham: Rowman and Littlefield.

Frankel, Mark S. 2003. Inheritable Genetic Modification and a New Brave World: Did Huxley have it wrong? *The Hasting Center Report* 33 (2): 31.

Harris, John. 2010. *Enhancing Evolution. The Ethical Case for Making Better People*. Princeton: Princeton University Press.

Holstein, James A. 1995. *The Active Interview*. London: Sage.

Hyden, Lars-Crister. 2008. Narratives in Illness: A Methodological Note. *Qualitative Sociology Review* IV (3): 49–59.

Kühler, Michael. 2013. Who Am I to Uphold Unrealizable Normative Claims? In *Autonomy and the Self* ed by Kühler, M. and N. Jelinek. Dordrecht: Springer.

Kühler, Michael. 2017. What if I Cannot Do What I Have to Do? Notions of Personal Practical Necessity and the Principle "Ought Implies Can". In *Dimensions of Practical Necessity. "Here I Stand. I Can Do No Other."* ed by Katharina Bauer, Somogy Varga, Corinna Mieth. Cham: Palgrave Macmillan.

Kühler, Michael, and Nadja Jelinek. 2013. *Autonomy and the Self*. Dordrecht: Springer.

Mehlman, M.J. 2009. *The Price of Perfection Individualism and Society in the Era of Biomedical Enhancement*. Baltimor: John Hopkins University Press.

Mitrović, Veselin L. 2015. Resilience: detecting vulnerability in marginal groups. *Disaster Prevention and Management* 24 (2): 185–200.

Mitrović, Veselin L. 2016. Parents' religious and secular perspectives on IVF planning in Serbia. *Journal for the Study of Religions and Ideologies* 43: 48–81.

Mitrović, Veselin L., Dónal P. O'Mathúna, Iskra A. Nola. 2019. Ethics and Floods: A Systematic Review. *Disaster Med Public Health Preparedness*. Available at: https://doi.org/10.1017/dmp.2018.154.

Universal Declaration on Bioethics and Human Rights. 2005. From the 33rd session of the General Conference of UNESCO 2005.

Veselin Mitrović is a senior research associate at the Institute of Social Sciences at the University of Belgrade. His research activities feature topics in Bioethics, Disaster Study, Sociology of Culture and Social Action in different strata (from marginal to elite). His recent publications include "Ethics and Floods: A Systematic Review." *Disaster Medicine and Public Health Preparedness*. 13/4, 2019, 817–828. "Parent's Religious and Secular Perspectives on IVF Planning in Serbia," *Journal for the Study of Religions and Ideologies* 15/43, Spring 2016, 48–81; "From Global Bioethics to Neuroethics," *TEME* XL/4, 2016, 1457–1475 (Serbian); "Resilience: detecting vulnerability in marginal groups," *Disaster Prevention and Management* 24/2, 2015, 185–200; "The Contingency of the 'Enhancement' Arguments: The Possible Transition From Ethical Debate to Social and Political Programs," *Journal for the Study of Religions and Ideologies* 13/37, Spring 2014, 93–124; "Ethos of the Elite and Future of Serbia," *Balkanske sinteze* 1/1, 2014, 167–189 (Serbian); "Nanotechnologies and Global Survival," *Sociologija* LIV/1, 2012, 37–54 (Serbian); "The Myth of Moral Enhancement: Back to the Future?," *Filozofija i društvo* XXIII/2, 2012, 111–123 (Serbian); "Human Enhancement: Toward the Creation of Patterns of Injustice?" Zeller, Jörg/Riis, Ole Preben/Nykänen, Hannes (eds.): *Issues in Theoretical and Applied Ethics*, Aalborg: Aalborg University Press, 2013; "Study," *Sociology in Serbia 1959–2009, Institutional Development*, 59–107, (ed.) Marija Bogdanović, (Beograd: Službeni glasnik i Filozofski fakultet Univerzitet u Beogradu, 2009) (Serbian). He is also author of the following books written in Serbian: *Apathetic Society* (2015); *The Stride of Bioethics, New Bio-Technologies and Social Aspects of the "Enhancement" of the Healthy* (2012) and *Jazz as Socio-Cultural Improvisation—A Qualitative Research of Social Mobility* (2012).

Index

M. Kühler and V. L. Mitrović (eds.), *Theories of the Self and Autonomy in Medical Ethics*, The International Library of Bioethics 83,
https://doi.org/10.1007/978-3-030-56703-3

The manufacturer's authorised representative in the EU is Springer Nature Customer Service Centre GmbH, Europaplatz 3, 69115 Heidelberg, Germany. If you have any concerns regarding our products, please contact ProductSafety@springernature.com

Printed and bound by CPI Group (UK) Ltd, Croydon, CR0 4YY
30/09/2026
02225696-0011